Derrick worked as a mechanical engineer before entering the fledgling computer industry in the 1960s. He worked in software development, becoming a senior manager in a large electronics company before becoming a trainer in a small partnership. He started up his own business in Melbourne, Australia where he now lives. He is now retired and has become an almost full-time adventurer and bushwalker.

Dedicated to my wife, Gina Hopkins, without whom I would be lost.

Derrick E. Brown

AN UNEXPECTED DEVELOPMENT

AUSTIN MACAULEY PUBLISHERS™
LONDON • CAMBRIDGE • NEW YORK • SHARJAH

Copyright © Derrick E. Brown (2018)

The right of Derrick E. Brown to be identified as author of this work has been asserted by him in accordance with section 77 and 78 of the Copyright, Designs and Patents Act 1988.

All rights reserved. No part of this publication may be reproduced, stored in a retrieval system, or transmitted in any form or by any means, electronic, mechanical, photocopying, recording, or otherwise, without the prior permission of the publishers.

Any person who commits any unauthorized act in relation to this publication may be liable to criminal prosecution and civil claims for damages.

Cover Photography: Copyrights of Tendring District Council

A CIP catalogue record for this title is available from the British Library.

ISBN 9781787100398 (Paperback)
ISBN 9781787100404 (E-Book)

www.austinmacauley.com

First Published (2018)
Austin Macauley Publishers Ltd.
25 Canada Square
Canary Wharf
London
E14 5LQ

ACKNOWLEDGEMENTS

I would like to thank many people who have helped me along the way and without whom I would be much poorer in all kinds of ways; my work colleagues who have helped me over many hurdles and from whom I learnt a great deal, in particular the late Colin Corder, ex-business partner, author and extraordinary man who died too young. Also Grahame Stehle and Marion Wells, ex-business partners, highly talented individuals and valued friends.

Thanks also to other colleagues from my two training companies and those many delegates, English, Australian and others who had to listen and put up with me on my training courses, who were so willing to share their experiences with me and with whom I had so much fun.

Finally the Scout Movement and the leaders from my Clacton Scout Group who helped me find myself and who had more confidence in me than my parents.

Contents

Prologue	11
Chapter 1	12
Early Days	
Chapter 2	23
School Days	
Chapter 3	47
Finding Myself	
Chapter 4	65
Early Career and Marriage	
Chapter 5	81
Sainsbury	
Chapter 6	122
Plessey	
Chapter 7	145
A Career Change	
Chapter 8	176
A Difficult Time	
Chapter 9	188
And Now For Something Different	
Postscript	206

PROLOGUE

Paris, 1973. A large conference centre. The chairman waited for the applause to end, then he announced the next speaker. I stepped up to the podium and looked out from the stage to the blur of almost five hundred faces. I placed my notes, written on 80-column punch cards, in front of me and adjusted the microphone. Somewhere there was a soundproofed room where a number of interpreters were waiting to interpret my talk into various languages. I looked up and began my introduction with my awful schoolboy accent "Bonjour mesdames et messieurs…"

CHAPTER 1
Early Days

I was born during the dark days of the Second World War, in June 1941 when, ten days after my birthday, Hitler's armies invaded the Soviet Union and Britain started to ration clothing. Greenford, in west London, just a few yards from the Western Avenue was where the birth took place, though whether it was at home or in a hospital I am ignorant. I was born on my mother's birthday and she apparently asked my father "Is this all I get?" Greenford's second claim to fame is the old Art Deco Hoover factory on that same avenue that is now a Tesco superstore. The Northolt aerodrome was just up the road, where, before I was born, Neville Chamberlain had returned from seeing Adolf Hitler and brandished the famous but soon-to-be proved laughable, Peace Declaration. My father was to be a part of the subsequent war and was conscripted, at thirty-three in 1943. My earliest memory is being at the front door, my mother crying as he walked away down the street in his army uniform, destined to take part in the greatest invasion force that the world has ever known.

Fortunately he returned safe and in one piece after serving in France, Belgium, Holland and Germany. He was in the REME (Royal Electrical & Mechanical Engineers) and drove a truck, towing a radar or a gun and he took part in the invasion of France on June 8th 1944, D day +2. He drove his truck onto the beach from a landing vehicle, having to follow marked lanes to avoid mines. The vehicle in front of him strayed from the lane and hit a mine. He'd been instructed to stop for nothing so he also had to drive to one side to avoid the smashed, burning vehicle. He had another narrow escape when his truck was hit by a train on a level crossing in Holland, killing his mate. The truck fell into a canal, and Dad was trapped in the cab as the doors were jammed. He escaped by putting his feet through the window and swimming out. He was in hospital for a short time before returning to the fray and to the job of beating Hitler. Fortunately he had lots of help.

Dad had poor teeth as a youth – early photographs show a big smiling face full of large teeth – and when he joined the army the dentist thought his teeth were so bad that he took them all out. Dad went to France without any teeth and had to live off hard tack biscuits for the first days as his unit was always on the move and had no proper food. He soaked the biscuits in his mug of tea and sucked them. His teeth eventually arrived by post and he was happy – they were a good fit. Then when his unit was camped in a field it came under attack and a shell landed nearby, blowing him up against a truck. When he came to, in hospital, he had no teeth. They were in a mug by his bedside, in pieces! He had to wait for weeks again for another set to be sent from England. I have this great admiration for the army system – in the midst of supplying hundreds of thousands of men, tanks, guns, ammunition and supplies across the English Channel, it was able to locate his pattern and despatch 'false teeth, one set, for the use of Private Brown, Alfred Joseph'!

He also was a despatch rider for a time and on one occasion he had to search for another despatch rider who had failed to turn up. Being warned of the danger of catch-wires he drove the motorbike with his head low over the handlebars. He found the missing motorbike and a headless corpse some distance up the road. The rider had been beheaded by the catch-wire strung across the road.

While all of this was going on I was presumably gurgling away in Greenford unaware of the dramas of the war. I have vague memories of seeing army vehicles on the Western Avenue, just a few yards from our house, but little else. The London Blitz occurred in 1940 to May 1941, ending just before I was born. Some twenty-eight thousand civilians died in this, and more of course died subsequently, bringing the total civilian deaths to about forty thousand. We had a bomb shelter in the house (Morrison shelter) – a large steel table affair with wire mesh sides that you were supposed to get under in bombing raids – I can't remember getting under it but apparently we did. This type of shelter proved to be effective in saving lives when houses collapsed, as many did from the blast effect of a bomb. They were of no use in a direct hit. There had been a large hole in the garden where my father had tried to install an Anderson air-raid shelter but the hole kept filling with water and he gave up. There were many air raids on London of course, and from June to October 1944 doodle-bugs (V-1 rocket) were used to target London. The V-1 rocket could be heard as a spluttering, buzzing drone. If you heard the engine cut out it then it was going to land and explode near, if not on you. The Northolt aerodrome was a target and some four thousand bombs landed within a two-mile radius of Northolt and Greenford caught some of these. The aerodrome was wonderfully camouflaged to look like a housing estate, so much so that returning planes sometimes couldn't find it! The V-1 rockets killed about six thousand Londoners. To minimise the effects of these the government 'leaked' information about the apparent landing places of the rockets. This information was actually kept secret but false data was leaked in a way that fooled the German planners into believing that the rockets were landing on the northern edge of London, so missing central London which was their target. So they shortened the flying time by ensuring the engines cut out earlier. The result was that many of the rockets landed in the lighter populated area of southern London, no doubt killing people there but missing the centre with its nerve centres of the military engine. The V-2 rocket was a ballistic missile that made no noise before it landed with devastating effect. It travelled at up to four times the speed of sound and there was no defence against them. They were used from September 1944 to March 1945 when the launching sites were overrun by the Allies. About two-thousand seven hundred Londoners died from the effects of these. My mother had to cope with a young family in the midst of this and with severe food rationing along with the worry that my father might not return.

I had an elder brother, Ron (born May, 1934) and sister, Jean, (born January, 1938). They were evacuated to the country to be out of danger from bombing but I was too young to be despatched. Jean proved to be a bit of a handful so both she and my brother were returned to take their chances in London. Another sister, Jennifer arrived after the war.

My father was demobbed from the army in August 1946. Until that time he was, presumably, helping get Germany back on its feet and was based in Hamburg. However only recently did something unusual come to light. In my late parents' belongings we found a beautifully-produced calendar for the year 1946, produced by a German woman named Helga. She included photographs of pre-war Hamburg and a text, describing her time over the latter part of 1945 with her 'beloved Alf'. She described how Dad was demobbed in August of that year, and took leave to go home, but that he returned to Germany and it wasn't until Christmas 1946 that he finally returned to England! I have been able to verify the demob date from the official war service record. There are many unanswered questions here. What was the nature of their relationship? What was he doing in Germany after his demob? Was he on some official war business or was he contemplating setting up with Helga? What did our mother know of this? Did our mother know of the existence of this diary? It would have been difficult to have kept it hidden all those years, but it was kept, and not destroyed. Were there any offspring? We'll probably never know the answers, but it sheds an interesting light on our parents' relationship. Of course, there would have been many such relationships between the Allied soldiers and the women of the occupied countries, even though British soldiers were forbidden to fraternise with German women. They were told, in a little booklet entitled "Germany" that was issued to each soldier in early 1945 that one in four German women had VD and that many of them would throw themselves at the soldiers in a hope of getting to Britain to a new life. Again, army regulations forbade marriage to a German, but some relationships did flourish and eventually lead to marriage.

On his return to civilian life Dad worked as a driver with the Brylcreem company on the North Circular road. He had a motorbike, a Royal Enfield (for some reason I remember the registration number – NVX 767) with sidecar. We had a family excursion on this contraption, and returning from Southend one night we had a puncture. With the tyre ruined, no spare and unable to get help, Dad stuffed the tyre with grass and we limped home very slowly. I remember standing by the side of the road watching him stuff the tyre.

My sister and I went to the nearby Selbourne School. I attended only for a short time before we moved away from London and I have little recollection of it. I notice that now it is a top-performing school! I just remember walking to and from home to school along the Western Avenue and over the footbridge. This road is now a major highway and you wouldn't let children anywhere near it. My first adventure was as a small baby in a pram. My mother had taken me shopping to the nearby shopping parade. Then it had proper shops like the butcher, baker, greengrocer and fishmonger. Now it has estate agents, a bookmaker, building society and banks. She went home and had afternoon tea with her mother, who lived next door with my grandfather and my auntie. A close family! "Where's the baby?" queried my Nan. "Haven't you got him?" responded my mother. I was retrieved from outside the butcher's shop. History doesn't record what I thought of this, maybe I was hoping to be collected by someone and taken to a life of luxury. I think that the butcher's shop window was imprinted on my tiny brain as I am a carnivore to the point of vegetables being only on my plate to add a bit of colour. Perhaps my mother was subconsciously trying to reduce the family size in the interest of survival of the rest, as food rationing was severe.

In 1947, within months of my father returning home from the war, the family moved house – to Clacton-on-Sea, Essex, to where my father's parents had moved.

My parents bought a guest house near to the seafront and started up in the holiday business. My maternal grandparents, Nan and Pop, also moved there. Clacton was in a mess at first, with barbed wire and gun emplacements still around, and the pier broken in two to prevent possible invasion troop landings. There were pill boxes on the beach and a couple of these remained, gradually subsiding into the sand until close to the turn of the century. With the returning troops looking for their first holidays it was important to open for business quickly and the first few seasons were very good. The seafront was cleaned up and the ice cream, candy floss, slot machines with 'What the butler saw' and 'kiss-me-quick' hats all appeared. We had deep snow that first winter and we had photographs of sledging and deep snow in all the streets. Clacton had had troops stationed there and the house we bought had been used as a soldiers' billet. It was three storeys, had eight bedrooms and a large cellar and we could soon take up to twenty-four guests. It was a corner house with a large garden. Dad built rooms that we called the annexe – a somewhat basic free-standing unit made from breeze blocks at the back of the house for the family to sleep in during the summer. This freed up rooms for the guests, although I ended up in a garden shed for some of the time or in a room that he'd originally built as a garage for the motorbike and bikes. Facilities then were not what we expect today. There were no en-suites – in fact there was only one bathroom with a bath, (no shower), washbasin and toilet, and one separate toilet, for all the guests. Washbasins were installed in all the bedrooms, but that was it. During the winter the family used the main bedrooms of course. Mine was no 1, at the top of the stairs and looked out at the back of the house.

I grew up in a house without any books, except for a 1930s set of the Encyclopaedia Brittanica which I dived into frequently. To my knowledge no-one else in the family ever looked at them. My parents read nothing at all, and they came from book-free households. I therefore did not experience bedtime stories as I had no story books, but I read comics. I therefore never got to 'Winnie-the-Pooh' and all the other famous children's books, so that when my daughter came along I read bedtime stories to her with, I hope, an extra dimension! I heard recently that my upbringing would be described now as a 'deprived childhood', but I have never thought of it as such. My first book was a school prize, and I then found the local public library and began, perhaps belatedly, with *Enid Blyton* and *Just William* stories. I slowly graduated to *Treasure Island* and boys' adventure stories. I discovered from my brother as he was near death that he had never read a book in his life! He said this with a kind of pride, and added that he had never seen the need for book reading. Again, following the family tradition! For their news my parents read only the *Daily Mirror* and *The News of the World* so their understanding of world events and politics was extremely limited. Their attitude to life was that of the "poor man at the gate" – a life of work, poor pay, subservience to authority – and ignorance. The passing on of information was almost medieval – news or more frequently (distorted) views would be passed on from the delivery men, the butcher, the knife sharpener, as if it were fact. No checking, no knowledge of how to check. Anything said by any authority was believed in entirety, so the BBC radio was listened to (but not that intently) and was almost the sole source of knowledge of events. When, as a paper delivery boy, I became aware of other newspapers, I realised what trashy papers they were reading, and I persuaded them to take another. They eventually agreed to take the *Daily Mail,* which was a tiny improvement. What I now think really strange was that they had no ambitions or expectations for their four children – there was never any discussion of what we might be or do. They believed that boys worked with their hands, while girls just got married. Professions were presumably for "other people", certainly not for us. But it was worse than that – I grew up believing that I couldn't and wouldn't be able to do anything worthwhile in my life, and it was no good thinking any differently as we just had to accept our position – at the bottom of the ladder. If ever I mentioned doing something or having something that was out of their experience I was told "Don't be silly. You can't/we don't/we can't afford…" and I believed it.

My father was a handyman around the house and could make a reasonable job of most carpentry, decorating, general repairs and plumbing work. I would watch and help so that I became reasonably good at this sort of thing myself when I married. However he certainly had no sense of taste or colour so we sometimes ended up with some rather hideous results. In the Fifties and Sixties hardboard became the flavour of the month so he boarded over rather nice panelled doors and made dreadful curtain pelmets.

We didn't have a TV until the Fifties, but I listened to the radio. I liked The Billy Cotton Band Show, Paul Temple, The Archie Andrews Show, Jewel and Warriss and of course the Goon Show. My mother listened to Mrs Dale's Diary which was always on as I arrived home from school and I wasn't allowed to talk until it was over, by which time I had lost the excitement of relaying the day's news. I shall never forget walking into the small living room on a wet winter's afternoon to be greeted by washing hung over every possible hook and piece of furniture and clothes horses spread around the coke-fired boiler. It is a pet hate of mine now to see washing drying indoors, except in a laundry room. The streets were lit by gas lights and a man cycled around with a long pole which he used to light each street lamp one-by-one. They gave a warm and rather romantic glow, especially in the snow. I notice that parts of London are still lit by gas lamps to this day. Horses and carts were still in use by rag-and-bone men and knife-sharpeners, green-grocers and gypsies selling wooden clothes pegs were all common door-knockers.

Looking back at my relationship with my parents I am somewhat surprised to find that we were really quite distant. I do not think of it as a warm and loving relationship either when I was small or later as a youth. It may have been a sign of the times but we didn't go out together as a family, I cannot even remember ever walking with my parents anywhere, and I'm sure that this was much the same for my siblings. So different to today's world where parents and children seem to be so much closer and love and affection are more obvious. What an improvement!

I am grateful to my parents for one thing in particular – a strong sense of a work ethic. I certainly grew up expecting to have to work for a living, in fact it seemed to me that there was a certain pride in the family in working hard for minimal wages. I never agreed with the minimal wages bit, but I certainly didn't expect handouts. Although I was unemployed for a short time in my working life I never wanted or received any government unemployment benefits, sickness or any other payouts of any kind. I've come across a number of people who expect and demand benefits and a number who have no intention of working if they can get away with it. I also had a strong sense of what is right and proper and this has been a great help to me in life when I've been confronted with decisions. This has got me into bother on a few occasions when I have spoken out against the tide as it were. More of this later.

CHAPTER 2
School Days

My mother's sister joined us at Clacton, taking a part in the business venture. My parents weren't particularly good businesspeople, my aunt was perhaps a little better. They bought the guest-house for £4,500 in 1947 and sold it for the same sum in 1960, when, it must be said, the golden days of the seaside resort were a memory. At first they charged ten shillings and sixpence for bed and breakfast (fully cooked), and had long discussions before eventually raising this to twelve shillings and sixpence! The main income though was from full board accommodation, (which was initially charged at around £5 per week) with many people coming back year after year for their week or fortnight by the sea.

The first few years were good, with visitors stuffing money into a pot kept on the hall sideboard for tips. Mum did the cooking. She was a good if plain cook and managed to cook for a full house plus the family without too much drama. The washing up and preparation was something else. We children would get roped in to help and of course we were reluctant. My sister Jean eventually negotiated payment to serve in the dining room and for general help. Brother Ron managed to keep out of it as far as I recall, but he strenuously denied this when I mentioned it! I got roped in as I became more capable. Breaking the crockery seemed to be one way to get thrown out of the kitchen but Mum became wise to that quickly. Auntie looked after the 'front-of-house'. She was pretty good with the customers and did the bedrooms and served in the dining room.

At first Dad was part of the team. This didn't seem to work well as he and Auntie fell out too often and Dad, I think, was not too comfortable doing this work. I recall him sitting outside the kitchen door, in his vest, peeling potatoes – bucketloads of them. After some time he returned to a driving job – on the buses. He drove for the Eastern National until he retired.

We had no mechanical kitchen aids, and we didn't even have a refrigerator. The large cellar that ran most of the length of the house kept things quite cool. If ice-cream was required then its purchase would have to be timed to the minute. When eggs were in short supply after the war they were kept fresh in the cellar in water glass (sodium silicate solution). There was only a limited variety of food available in the late 1940s and food rationing persisted until 1954 when finally meat was de-rationed so it must have been quite difficult to plan the menus.

Clacton was a famous seaside town in the Fifties, with its second-longest pier in the country, a large Butlin's holiday camp and many hotels and entertainments. Cliff Richard, Frankie Howerd, Roy Hudd and many other entertainers started their professional lives at Clacton's

Butlin's as 'Redcoats'. It may be difficult now to see what the attraction was in staying at Butlin's. A small chalet, eating in huge dining rooms, being with crowds all the time... maybe this was very acceptable at first, just after wartime, but the attraction waned. One big attraction was the entertainment, all laid on, all free. For the single young people there was another attraction – sex. Without parents around, young men and women were able to do what they wanted, and the chalets were arranged to assist in this, with alternate rows of male/female chalets, and again chalets for opposite sexes above/below each other. Eventually the seaside holiday scene all began to go downhill as the public became more sophisticated and wanted more. The cheap Spanish holiday packages became a serious competitor. Slowly the hotels closed and were pulled down or became retirement homes. The Clacton Butlin's closed in 1983, the guest houses became cheap bedsitters and recently I was sorry to see our old house looking very shabby and rundown, with rubbish in the front garden. It had always been so bright and smart. The rear garden had been sold and four apartments built there.

I went to school at the Holland Road Primary school, going by bus – a penny-halfpenny fare. I did reasonably well, but not brilliantly, always coming in the top half of the class. The school had been used by the military and was under-resourced at first and became overcrowded – the classes were large by today's standards – the year I took the eleven-plus exam (1952) there were fifty five children in the class. It seems that today's teachers can't handle thirty!

These early schooldays were largely uneventful. I managed to break an arm while playing leapfrog, and the head teacher, Miss Ault, took me to the hospital in her tiny

Austin Seven. She was the only teacher who drove to school – cars were not commonplace then. I had broken both the ulna and radius of my left arm and proudly wore the plaster-of-paris splint for some weeks. At some point I won a prize for being in a team that came second in the tug-o'-war! I was also awarded a book prize for something – I forget what – this was "Kidnapped" and was one of the first books I ever owned. This was the catalyst that started me reading. I was laid low by a cricket ball that thumped me on the collarbone when I was in the (very) silly-mid-off position. In the last year there I won a swimming prize and one for the best boy country dancer! We had a lovely teacher, Miss Teagle, and with her we formed a Morris dance team and performed at events dancing as one sees them today in English pubs, complete with bells on legs and head bands. We also danced around the Maypole weaving patterns with the ribbons and we even managed to unweave them sometimes! We were never told the origins of the Maypole though, the phallic symbol of spring. I have a photograph of myself in the sword dance team, holding up the swords made into a pattern. At some point I boxed, winning my two fights on points but I think that neither of us landed any good punches.

In those days you took the eleven-plus exam and that decided which secondary school you went to. The day the results arrived Dad opened the mail and told me I'd failed. He was disappointed, but he told me as if he'd expected nothing different. At school I was told that I'd passed; the letter just informed me that I needed to take an interview as well! Either Dad had misread the letter or it was badly-written. I think the latter, as many children were in my position. After the interview, that was rather daunting, I was in and started at Clacton County High School in September

1952. Not one of my old school friends was in my new class, although there were several in the same year (there were three classes for each year). It was a grammar school and much more formal than the junior school. Each morning commenced with assembly, with the headmaster, in mortar board and gown, rather pompously walking up to the stage in a hushed atmosphere where he would conduct proceedings. He was a remote figure, and I never exchanged a single word with him in my seven years at the school, and this included my two years in the sixth form, when there were only about twenty of us. What the heck was this man doing if he didn't even speak to his sixth formers?

The grammar school was a whole new world for me – a uniform with cap, blazer, games clothes, the lot. We had gym and games lessons and after these we had showers. This was my first introduction to these as we'd only had a bath at home. Showers were only slowly to become popular with the British! I was the first of the family to be experiencing a grammar school, and I certainly felt that I was stepping out of line. I became conscious – and embarrassed – with my speech. My parents were near Cockney and our family spoke rather sloppily with, I soon realised, poor grammar. I was pulled up sharply when I said "I done". I had never realised that this was incorrect. My parents were very careful with money and I was made aware of the cost of my schooling. I got the minimum that it was possible to get away with. After a time as I had to renew things and pay for extras, I tended to pay for them myself as I began to earn money with after-school jobs. I badly wanted three things – a bike, a wristwatch and leather gauntlets. All the boys had these, it seemed and I was envious. Dad bought me a bike at Christmas when I was twelve but I had to pay for half of it

(£6). I bought the gauntlets, and I found a broken watch in the gutter and had it fixed.

I found the schoolwork strange. The idea of homework was new, and at first I was unaware of the need for it. I must have been dreaming my way through classes, because I was put into detention on a number of occasions for not doing the homework, and I was somewhat surprised that it was expected. Mum and Dad did not understand this either, and discouraged me from doing it, "It's wrong to expect this of kids," Mum would say. I got the hang of it at last. However, a competitor for my time had appeared on the scene – television. After the coronation of 1952 TV sets began to get more common. We got one in the mid-Fifties. I soon discovered that there wasn't a lot worth watching (at least on the channels that we watched) so after a short time it didn't bother me. However, Mum didn't think it right that I was working in my bedroom and kept on at me to join the family in the living room, where 'Take your Pick' 'Open the box' or something similar was on.

The coronation of Queen Elizabeth 11 was celebrated at Edith Road in June 1952 with flags and bunting. Dad fixed the Union Jack and a picture of the Queen across the second floor balcony and had spotlights in the front garden trained on the balcony like searchlights. I took a trip to London to watch the coronation – on TV. I stayed with the Osborne family at Osterley. They were ex-neighbours of ours from Greenford and the families had kept in touch.

On a fearsome night of January 31, 1953 the North Sea flooded the east coast and thirty five people were drowned locally. Nearby Jaywick was under water and many of the wooden bungalows were up to their roofs in water. I went to investigate with a mate and we walked along a wrecked seafront to Butlin's. The sea wall and promenade had been ripped apart as if made from polystyrene. All the funfair was flooded, the ghost train was ripped open and the once-scary skeletons and howling faces now looked pathetic and smashed. We continued up the road to Jaywick where the waterline ended at the Three Jays pub where we saw the police and ambulancemen pulling a body from the water. One of my friends who lived there had to swim out in the dark and he bore a scar where he ripped his thigh badly on a barbed wire fence. For years after you could see the water mark near the ceiling of many houses. In all some thirty thousand people were evacuated and three hundred and seven died along the east coast that night.

My maternal grandparents, Nan and Pop, lived not far away, and I developed a rather special relationship with Nan. She had a great deal of common-sense and I would often visit her after school. We had great chats about everything, she taught me to play cards and she introduced me to the concept of being an entrepreneur. She had a number of apple trees that produced a good crop, far too many for her, or the family's requircments. She suggested that I pick the apples and sell them to the guest houses in our area. I readily agreed to this. I paid her threepence per pound for the eating apples and fourpence a pound for the cookers (there were 240 old pence to the pound then). I took samples around to our neighbours and took orders, selling them for double the cost price. I cycled home carrying large bags of apples on my bike and delivered them. So this was my introduction to running a business! It worked for a while but I realised that it took time and I wasn't earning much. I had also started to earn money as a barrow-boy and this paid a little better. Visitors arriving at Clacton by coach (there were many without cars then) would have suitcases and would either have to struggle with them to their hotel or guest house or be taken by taxi. The alternative was to hire a young lad with a barrow who would guide them to their destination. Dad built me a barrow using some old pram wheels and I would wait at the coach station for the London and Midland coaches to arrive on a Saturday morning. 'Carry your luggage, sir?' was the cry. I'd get paid one or two shillings (five/ten new pence) or a half-crown (two shillings and sixpence – 12.5 new pence), if I was lucky. I could earn several pounds on a good day. I gave this up after a couple of seasons when I found other work.

Both my brother Ron and sister Jean had done paper rounds so it was almost obligatory for me to take on one. This earned the princely sum of twelve shillings and sixpence per week – two and sixpence per day (I didn't do Sundays). If you got up at 5.30 a.m and collected the papers from the railway station you got an extra shilling a day, so after a time I did that as well. For two evenings after school I did a butcher's round – delivering meat on a trade bicycle. I did the same on Saturday mornings and this paid me another twelve shillings and sixpence. I did this for several years and graduated to occasional working in the shop making sausages, cutting up and jointing meat, even splitting pigs and lambs in two from head to tail. One Christmas Eve I was working on into the evening as the last delivery of turkeys had arrived from the farm, dead and plucked, but still warm. My job was to complete plucking and 'dress' the bird, removing the head, feet and guts. I was by myself working at a bench where the turkeys were piled high. I was cleaning (gutting) a turkey when there was a noise from one of them and it rolled off the pile! I was so startled that I almost fell into the bin of turkey entrails! My first thought, of course, was that a turkey was still alive. However, all that had happened was that the wings of one, tucked behind the back, had opened out and the lungs had sucked in some air!

It was whilst I was delivering meat one Saturday that I had my "Mrs Robinson" moment. The customers would often pay me for the delivery so I would collect the money, sign the invoice and make a record of this in a notebook. Sometimes I would receive a small tip. A new customer appeared in one of the better houses, and rumour had it that the couple had won the football pools. Certainly, a new Jaguar was often parked in the drive. All went normally for the first six weeks or so. One morning the Jaguar was absent and my door-knock was answered by the woman dressed in a negligee. I was invited into the kitchen while she found her purse to pay me. She had once been pretty but she had now run to fat, and her make-up was overdone and badly applied, I noticed the smeared lipstick in particular. A gin bottle on the kitchen table confirmed that the cornflakes had had more than milk with them. It was a warm day and she offered me a drink of lemonade. She took her time and engaged me in conversation. She passed me the money, leaning towards me. Her negligee gaped open, revealing quite a lot of what was supposed to be inside it. "You're a nice-looking young man," she purred, "I expect all the girls are after you." I wasn't sure how to answer this and mumbled something in response. "I wonder," she went on, "if you could help me with something upstairs. The wardrobe door is stuck and I can't open it." Now I was around fourteen or fifteen and sexually inexperienced. I was certainly interested in girls and had been having dreams of how I would lose my virginity, and the options hadn't included what was unfolding here. It took me all of a few seconds to decide that this was not what I wanted. I fled. I told no-one of this, not a soul, ever – until now. On subsequent visits to her I refused to cross the threshold and got away as quickly as possible.

Then I discovered a better way of earning money. During the summer I worked on stalls at the Butlin's fairground, eventually ending at a stall selling doughnuts. For a time I was juggling all of these jobs at once. Eventually I dropped all the jobs except the doughnut selling, and I kept this right through until I left school. The fairground experiences and the stallholders were fascinating. I found out how the coconuts were fixed so that it was near impossible to knock them down and how the 'pick a lucky number' systems were rigged. A woman would buy a straw that might contain a name and a number – only a straw with both would win – and she would extract a dummy straw from her pocket that would indeed contain the name and number that won the top prize. Much of this was made in the spruiking, drawing a large crowd, and she would take a long time deciding what prize to take – the giant panda or the twenty-four-piece tea set? She then walked off with the prize leaving an eager throng around the stall all anxious to spend their money. All they got was a useless empty straw, for that indeed was what nearly all of the punters got. A few of them won a cheap plaster duck or similar, something we called "penny trash", just to keep up some hope of winning something worthwhile. Late at night when the funfair closed all the prizes 'won' by this woman were returned to the stall to be 'won' again the next day. I know this because I worked on this stall. One day was enough for me, even though I was quite well paid, as I didn't want to be a part of this. During some of my time spent in the fairground I learned the sequence of a set of gambling machines. I was able to watch the sequence, then jump in with a few coppers and clean up by picking all the winners. The owners soon realised that their machines were not making them so much money and banned me, detailing a bouncer to throw me out, but I would loiter around until the coast was clear then nip in and win again. Eventually the machines became more sophisticated – the mechanism was randomised, which stopped my little racket.

I was fascinated by the 'chat' that the stallholders would use to entice customers to put their hands in their pockets. Some of them were highly skilled, not to say cunning, and wouldn't hesitate to get small children to spend their last pennies. The funfair experience taught me something of selling, that came in useful later. I was always cynical about funfairs as an adult, and this made me a dull companion and father to my daughter later in life when I took her to any funfair. I also learned that people loved to watch a demonstration and that they liked to join a queue. This appears to be a British thing, possibly connected to the need to queue for everything during the war. The trick was to always have someone at the stall – with no-one waiting to be served people would walk on by. As soon as there was one person at the stall, another would join. I learnt to chat to them, to adjust the machine, to sugar the donuts skilfully… and they would be content to wait. The queue would grow. Sometimes it would snake backwards and forwards across the promenade and people would be waiting for twenty minutes to get to the front. We could produce thirty two donuts a minute with two machines going.

The donut stall (that was the spelling – it was an American machine that made ring donuts at the rate of sixteen per minute) also provided a way of watching the holiday girls strolling by on the seafront. In the quieter moments I would try and chat them up and make dates for when I finished. This was often latish – maybe 10 p.m. or so in the high season so the owner, Buster, for that was his nickname, would let me off early when occasion demanded. In the high season we did very well and I would work the school holidays twelve hours a day for seven days a week during the key weeks of the season. In this way I saved enough to buy my first vehicle for cash– a Lambretta motor scooter – as soon as I left school.

Buster was a character. He was short, heavily built, and very jolly. He came from Great Bentley, a village just five miles out of Clacton, known for having the largest village green in England. He had been the village strong man, and his party piece was to lift a man under each arm. In fact, his arms were so short, and his chest so large that he could not fold his arms! He and I got on very well together and he offered to sell me the business the year I left school. He wanted to retire and he didn't feel that his son was capable of running it. I declined the offer, after a little thought.

Living by the sea was, in retrospect, a source of joy for me. In school holidays I'd spend much time on the beach and swimming. I became a confident swimmer in all but the roughest sea. I was a little overconfident when a very rough sea picked me up and threw me onto the beach like a discarded fish. I landed face down on the shingle and was dragged back by the retreating wave. I suffered quite badly from shingle scraping, looking like I'd been rubbed over with a cheese grater. As my parents were working and as our house was only 100 metres from the beach it was easy to stroll down for a swim at any time. The North Sea is never particularly warm, but one gets used to it. To this day I have no trouble swimming in creeks and pools in Australia that my friends refuse to swim in. Around the age of fourteen I bought a canoe from a mate for £1. It was a canvas-covered two-seater kayak that had been wrecked against a breakwater. I removed the canvas and repaired the frame, then recovered and painted it. Not knowing anything about canoe construction I had to find out as I went along. Marine plywood I established, was the wood to use for the frame. I used deckchair canvas to cover it, suitably treated to shrink it tight before painting. I did this by myself and was very proud of the result. With a trolley I could take it to the beach and launch it quite easily. It handled well, was quite stable and was quite fast. With some practice I could manage reasonable waves but catching a wave side-on could capsize the craft without trouble. It was nigh impossible to get back into the cockpit in deep water so I'd have to push it back to the beach to get back in.

When we canoe on rivers these days we wear life vests and helmets but then there was no thought of danger. Life vests, floatation devices and the like were never considered and I would often canoe a long way out from the beach. On one occasion I was caught in a swift current that threatened to take me under the pier. I knew that there were many broken, sharp-pointed stumps of old supports under the surface (from when the pier had been broken in two in war time) so it was not wise to pass beneath the pier in a canvas canoe! I tried to paddle away but after five minutes of fierce, not to say frantic, paddling I was in the same spot. Reluctantly I swung the canoe around and whizzed under the pier, holding my breath. I missed the stumps!

Canoeing, sun baking and swimming became a way of life in all spare moments during the warmer weather, when I wasn't working. It was also a distraction, so that when I was supposed to be preparing for my 'O' level GCE exams I was on the beach with my mates – and our canoes. I was discovering girls around this time and in particular the student teachers at the teacher training college that occupied what had been several large old hotels on the sea-front. Our chat-up line was to invite them to join us for a canoe trip – it worked well and we'd soon be out with them in our canoes. Once on a rough sea with a bikini-clad girl in my canoe we were taking water as the waves hit us. The canoe got lower in the water and it was becoming difficult to handle. Eventually we capsized, still more than a hundred yards out from the beach. I was trying to save the now upside-down canoe that was almost under the water, when a scream from my companion got my attention. She'd lost her bikini top. At that moment I was more focused on saving the canoe. We swam, pushing the canoe, to the beach where a crowd of holidaymakers had gathered on the promenade to watch the entertainment. They applauded as I got the girl out of the water, minus bikini top, then I got to the canoe. In righting it, the weight of water split the side from end-to-end, necessitating considerable repair work. My companion wasn't too happy, inferring that my seamanship wasn't perhaps of the highest quality. She was no doubt correct, but the relationship wasn't damaged too much as I dated the girl for some time after. Another relationship, albeit short-lived, was with a girl whom I took to a quiet beach to go swimming, hoping that I could perhaps advance my limited sexual experience another notch. I noticed that she had six toes on one foot. I asked her, optimistically, did she have any other extra features that I should know about?

Just around the corner from our house was the Westcliff Theatre. During the summer season it would run a revue with the usual sorts of acts, a compere and a troupe of showgirls. We had an advertisement board on our garden fence, that directed people to the theatre. We were paid for this with two complimentary tickets per week. I would often use this and either alone or with a friend I'd sit in the stalls enjoying the show. Some performers went on to become household names, while most disappeared without trace. One regular act featured "Bunny Baron – five feet of fun and a banjo". At some point during a quiet season we took some of the chorus showgirls as guests at the house. I became friendly with one (Wow! How did I manage that?) and was fascinated watching her in her skimpy costume and long legs in the dance routines. I secretly stole up to her bedroom on a couple of occasions when she returned from a show with all the greasepaint on and she would tease me as she washed it all off at the washbasin. My offer to wash her down was refused but merely watching her was almost as exciting! Our age gap meant that the relationship was going nowhere but I was thrilled to just be with her.

At the High School I was in the bigger pond with the brighter kids so I wasn't doing as well in class. I was probably a bit tired as well. By the time I got to school after the paper round I was ready for a sleep. The school put much emphasis on languages and I did not excel at the only one I was studying – French. I found it incredibly frustrating as I just could not understand the spoken word, and I would fail the dictation tests every time. Aural learning was not for me, but it was not until much later in life that I found out that we learn in different ways and have natural preferences. I did seem to find it difficult to follow and retain the spoken word, finding it much easier to learn by reading and doing. So I would have benefited by solo studying, but I was reluctant to do much of that. I joined the 'C' stream in the second and third years, where I was in the top five in class, coming top one term. To my surprise now, I find that I came top of the class in French, twice, according to my school reports. So I must have learned something! I did eventually scrape an O-level GCE pass in this subject, doing badly in the oral exam. Then I went to the science stream. I quite enjoyed the sciences. For the first two years we also did woodwork, and had to walk to the workshops at the youth centre in the town. Again I enjoyed this and made a letter rack and a coffee table. The coffee table is still around – with my younger sister Jennifer.

I was no great shakes in sports, although I enjoyed playing football. Cricket I never particularly enjoyed. I could sprint and did well at the 100 yards and 220 yards. We would also go cross-country running in all weathers and would run several miles up to Holland-on-Sea, running through ditches and muddy fields in the middle of winter. I wasn't very good at this, although I ran in the Scouts' Boxing Day marathons for several years. The school had a good gym and I enjoyed the gym lessons but again never excelled. In the summer we'd cycle or jog down to the beach for swimming. I could swim reasonably well and swam in the school swimming gala, held in the public swimming pool on the pier. Towards the end of my school career I became involved in the school dramatic society. I played a mad vicar in a house play, 'Ivy Cottage' and I also worked backstage for a school play, doing the props and lights. I enjoyed 'treading the boards,' finding it an exhilarating challenge to speak before an audience. Although it was somewhat nerve-racking, I found that I was able to speak well. Little did I know that I would end up doing this for a living!

Mum and Dad were 'religious' according to mum. That meant that they thought we should all be baptised, be married in church and have a church funeral. Whether she actually believed in a God I don't know. My father had nothing to say on the matter. As a girl mum had gone to Sunday school and she thought her children should attend. As far as I know she never prayed and she never went to church unless it was for the aforementioned three activities. I was made to go to Sunday school at Christ Church Congregational Church. I even appeared as missionary Albert Schweitzer on stage in a play there. I soon began to doubt the wisdom of spending time at Sunday school learning about what was essentially, mostly nonsense, in my view. And I was learning at school about science, about logic, about thinking for yourself. And yet, here were a bunch of adults very seriously taking the view that some magic had and did regularly take place. All we had to do was to pray to this being in the sky. The words of the hymns seemed so ridiculous – 'the gentle rain from Heaven' were in contrast to the floods that drowned many people, I thought. I remember my indignation that I had to be told what was right and wrong, and the parable of the shepherd looking after the flock – us! I did not want to be compared to a brainless sheep!

I began to seriously doubt all of this. The more I learned about different religions, about Catholicism, Islam, Buddhism and all the others, the more I became convinced that some people were pulling a con trick and that they had a great advantage to be gained if the con trick succeeded. Although I couldn't have expressed myself at the time like this, by the time I was eleven or twelve these ideas were taking shape. My school reports said "Only fair" for Scripture, reflecting my lack of interest and I began to resist Sunday school. My parents insisted that I go. In the end my sister Jean was instructed to escort me after I had refused. On the way I ran away from her and went to the beach. Jean didn't catch me or she didn't try. I never went to Sunday school again and I eventually realised that I was an atheist, and have remained one.

The family expectation was that I would leave school at sixteen, at the end of the fifth year, having taken 'O'-level GCE. It slowly became apparent to me that there was another option. I could stay on in the sixth form for two years to take 'A'-level GCE. If one passed this exalted exam with sufficient subjects there was the possibility of university. Now this was completely outside the culture of my family. We were working class and proud of it. My parents had left school at fourteen, were apolitical, never voted ('our vote doesn't count', my mother would say), read nothing except 'Woman's Weekly' and dismissed anything not in their immediate area of knowledge as 'rubbish'. This included opera, ballet, classical music, theatre… and certainly education beyond the age of sixteen. It was not until the fifth year that I raised the subject of staying on at school. They had no idea about university education and certainly no respect for knowledge. Only the "toffs" went to university in their world, but they didn't understand why they – or anybody – went there. Dad went crazy. I was called a lazy good-for-nothing, who just wanted to hang around, when the right and proper thing to do was to go to work and earn money. By now I'd seen the light, though I must say that my teachers generally were not much help to me. They were focussed on the brightest children who were going to win scholarships to Oxford and Cambridge and thus bring honour to the school. I was not in this category. There was absolutely no mentoring or individual help provided at all from the school and I was left floundering.

I was desperate and trying to find ways to stay on and to find ways of earning money to pay my way (as this seemed to be the major obstacle). All seemed lost and the end of the GCE exams and what looked like being the last school holidays for me drew near. Then out of the blue came Uncle Bill from London on a visit. He was Nan's brother. He was a great big fellow, full of swearing, bounce and fun. The important thing for me at that time was that he was an ardent trade unionist – a shop steward working for London Transport. When he asked after me and heard the tale from Dad he turned on my father. He called him all the names imaginable and told him that education was everything, and that I should stay on at school. This was news to Dad, who thought that he'd got this worked out. However, he listened and to his credit agreed there and then to allow me to stay on. I often wonder what my life would have been like if Uncle Bill had not turned up that day.

So, stay on I did. I had been buying most of my own clothes for years by now so the only cost to Mum and Dad was food. I didn't hear many complaints after this. I studied applied maths, pure maths and physics and got two 'A'-levels. I missed out on the third – pure maths, which I'd always struggled with. The way was now clear for me to go further.

When it was my younger sister Jennifer's turn at age eleven she was able to go to the same high school without drama. She became a bit of a rebel and received little support from Mum and Dad who had no idea how to handle her during her adolescent years. I was living away from home at this time so I was unable to help, but I could see that she was wasting her opportunity. I felt that I could have been an influence on her. My mother complained to me at one stage that she didn't do as she was told and she kept asking questions! She would ask "why" all the time! I tried to explain that this was natural, that it showed an enquiring mind and that it should be encouraged. Mum clearly thought that Jennifer should accept what they said without question. They really hadn't moved on from "children should be seen and not heard", that mantra from Victorian times. Jennifer stayed with me for a brief holiday when I was married and living in Luton. She was obviously bright and could have done well at school but this was not to be the case. She left school at sixteen.

During my childhood I'd been a Cub and a Scout and this was a source of pleasure and exploration for me. My brother Ron had been a Scout and I joined the same troop – the 1st Clacton-on-Sea troop. This, together with school, opened my eyes to new worlds. I was reading almost anything I could get my hands on. Although I had few books at home, I had joined the public library and we had a reasonable school library. At Scouts I was introduced to the outdoors and camping and this shaped the future for me to a great extent. I loved the camping and weekend walking. Carrying my gear, cooking over a fire, sleeping in a tent, being close to nature – all this had great appeal for me and still does to this day. I became a patrol leader and this was an early lesson in leadership.

At home I was never expected to perform or excel at anything. I'm sure it was the same for Ron and Jean as well.

Even worse for Jean –"education is wasted on girls" – Mum would say. In fact, had I said for example that I wanted to be a doctor, a schoolteacher, or any kind of professional, my parents would have laughed and said that that sort of thing wasn't for us. It was just outside their thinking. But I didn't say it – I was so brainwashed that the mere *thought* was outside my thinking. When the group Scoutmaster said that he expected me to earn a Queen's Scout award, I just about fell over. I had never considered this. I remember cycling home that evening realising that it *was* a possibility. With the First Class badge in my sights I realised that I could, with a bit of effort, work for the necessary further badges and get the Queen's Scout award. You had to earn this before your eighteenth birthday and I just managed this. So, in September 1959 I was presented with the award and on looking at the details now I find that I was only the fourth Scout in the 1st Clacton Group to achieve this since the war. Some months later I was presented to the Chief Scout at Gilwell Park at a weekend camp celebration. This was a significant step for me – I was breaking the mould in which my parents had placed me – the mould of unimportance, non-achieving, non-entity. I was finding that I *could* achieve something. It had just taken a word or two of encouragement to wake me up.

It was in the Scouts that I went abroad for the first time. This was an adventure for me when, at the age of fifteen, I went on a two-week walking/ camping trip to Germany. With my two mates, Ron and Peter we walked and camped around the Lahn, Moselle and Rhine valleys. We travelled with and met up with other senior Scouts at Koblenz but we hiked in our team of three. One evening we decided to do a night hike. By the early hours we'd had enough and wanted to make camp. It was wet and misty and in the dark we couldn't see much at all but we found what seemed to be a suitable grassy site. When I awoke next morning the first thing I saw was a huge cross with a crucified Jesus looking down at me! We were camping in a graveyard! We moved on rather quickly. We called in on a beer and wine festival and of course we took advantage of the free wine and beer tasting. This was all a great adventure.

I had my first taste of the mountains when I went to Snowdonia with a senior Scout expedition in January 1959 when we climbed Mount Snowdon and camped out near the summit. It snowed heavily during the night and I awoke to find the tent bearing down on me under the weight of snow. We dug our way out and saw the most fantastic sight – the sun glittering off the pure snow that was banked up all around us. Such a thrill! We also scrambled along the notorious Crib Goch ridge on the Snowdon horseshoe roped together as the rocks were covered in ice and snow. I found all of this most exciting but it was only much later in life that I was to do much more mountaineering.

Another expedition with the senior Scouts was to explore an underground system of caves. These were the Dene-holes, or Dane holes and were manmade. They were in Hangmans Wood at Grays, Essex and there had originally been sixty or more vertical shafts some sixty feet deep, each having four chambers symmetrically arranged at the bottom in a daisy-leaf pattern. Many shafts had been filled in but connecting tunnels between chambers had been dug so that a veritable rabbit warren now existed. Originally the chambers were not connected. They were pre-Roman, but their origins were unknown. There were several theories as to their purpose, the most likely being either storage of grain, or for the mining of chalk, for fertiliser. If the latter, then it was odd that they were not connected. I favour the storage theory, as it would make sense to keep them separate for security reasons. When invaders were around (and being near the River Thames Grays would have seen many invaders) the locals would have wanted to hide their grain, maybe valuables and themselves down there as well and cover the tops of the shafts. Even if one had been discovered, there remained many others. We turned over these thoughts as six of us descended using mountaineering ladders and spent two days down there, exploring all the chambers and tunnels, and slept down there overnight. I was the first one to descend and it was a little scary going into the unknown. The circle of light that was the shaft entrance became smaller and smaller as I climbed down until there was pitch blackness at the bottom, when the ladder suddenly swayed as I entered the chamber. I landed on a soft pile of debris, and released the safety line which snaked back up the shaft. I waited by the ladder as the others climbed down and our rucksacks were lowered containing sleeping bags, cookers, food and water. We explored the chambers and tunnels thoroughly, some of the tunnels being quite tiny necessitating us to crawl through with our climbing helmets bumping against the roof. We found no artefacts, and little or nothing was ever found when the site was officially excavated. Our night passed without event and we returned to the surface next morning feeling very satisfied with our little adventure. Apparently now there is only one shaft left, the others having been filled in, and permission is required to access the chambers. In those days we just went there and climbed down without even telling anyone!

I also attended the World Jubilee Jamboree at Sutton Coldfield near Birmingham in 1957 with thirty five thousand international Scouts. All in all I have much to thank the Scout movement for. In particular three brothers gave up a huge amount of their free time to run the Scout group and they played a large role in encouraging me. I left the Scouts at eighteen when I went to college and I always meant to give something back by perhaps becoming a Scoutmaster but to my regret, I never did.

CHAPTER 3
Finding Myself

Summer holidays with the family didn't exist in my childhood as summer was the time when we all worked. Winter at Clacton was the quiet time but of course this wasn't convenient for holidays – at least in the UK – and no-where else was ever on the agenda. So we never went away together. However, living by the sea had its compensations as I saw more of beaches and swimming than most children. There was one area though that with hindsight I feel that I missed out on. We never ate out. I remember being taken to a restaurant on Clacton sea-front as a treat for a cup of coffee and a cake. This was something very special – I was overawed by the place and the service. I grew up in awe of waiters and was not at all confident with the cutlery or with ordering. It was not until I was in the sixth form at school that I attended a formal dinner and I was embarrassed not knowing which piece of cutlery to use. I had an issue with taxi cabs as well. It had been drummed into me that these were prohibitively expensive, so that when I later lived in London and could afford the odd taxi, I was extremely nervous about using them. This feeling persisted for some time, even though the cost was of little consequence to me.

There was one area that held no fear for me – public speaking. I didn't mind reading the lesson at church (and I well remember reading my first lesson on a St George's Day Scout parade), taking a role in a play or making a speech. I was nervous, sure, but I seemed to enjoy the challenge. I appeared on stage in the Scout Gang Show at Clacton's Savoy Theatre and enjoyed playing roles in a couple of skits so it's not so surprising then that later in life I earned my living as a speaker and presenter.

As my schooldays came to a close I had no idea what I was to do. I'd always had a vague idea of being an engineer and I'd shown some mechanical aptitude with a Meccano set, but that was all. I did want to travel, and that came about, eventually. My two 'A'-levels were insufficient to earn me a place at university, the provisional place that I had depended on three passes. The Hoffmann Manufacturing Company at Chelmsford stepped in and offered me a position as a graduate engineering apprentice. This involved a 'thick' sandwich course – six months of each year at college and six months in industry. The college was Enfield Technical in north London. The company made ball and roller bearings and they provided a solid training programme. What's more they paid me to study, and paid an allowance for books. The starting pay was £8 16 shillings (£8.80) a week, which increased annually. This was more than most of my friends received from study grants, and I was paid in holiday times as well. After the first six months at college I had a week's holiday before starting the first industrial-period at Chelmsford. My pay arrived by registered post as I lay in bed one morning. My mother saw my payslip and was horrified.

"Good grief son," she cried, "Here you are lying in bed and you're earning as much as your father!" I obviously shouldn't have been paid this much. Our lot in life was to work for little pay. I had had no idea until then that bus drivers were paid so little.

So I started work for the first time in my life. I had to get up at 5.30 a.m. to get to the railway station and catch a 6.20 a.m train so that I could arrive at the factory at the appointed 7.30 a.m. start time. We finished at 5 p.m. so I didn't arrive home until almost 7 p.m. I started by learning to use basic machine tools (I still have the tools that I made in the apprentice school) then in the subsequent industrial periods I had spells on the production lines, the work study, drawing office, and technical advice departments. I enjoyed these experiences and found my way around the huge factory site that employed about five thousand people. It had its own power station, foundry, and had a worldwide reputation for high quality bearings. One of my duties was to act as a guide to visitors and I would sometimes take parties of schoolchildren around. They were often fascinated by the shiny steel balls and the bearing parts and would slip some of them into their pockets when they thought that I wasn't looking. As we were having the customary tea and cake on one occasion a teacher noticed the bulging pockets and made the boys empty them out on the table. I had to dispose of the booty – it all went for scrap as it was now useless for production!

I spent six months on the production lines – vast areas and terribly noisy. The air was thick with water and machine suds droplets, so that the production line seemed to disappear into the horizon in a thick haze. I would work with an operator on a particular machine, taking notes and learning how to use it. I was free to move on to the next machine lined up for me at my choosing. Sometimes I was ready to move on after a week. To talk to someone you had to shout into the other fellow's ear. I sometimes wonder whether my deafness later in life was due at all to this environment. Individual machines were powered by belts driven by overhead shafts. These belts were the cause of some horrific accidents when operators were caught up in a broken belt, one chap being taken up to the factory ceiling and killed. Although old-fashioned, we were told that the company kept them to demonstrate the use of the bearings! I now doubt that this was true, it was more likely that the place was under-capitalised. I was amazed at the age of some of the machines – one grinding machine in regular use was dated 1896! The old fellow working it was a real craftsman, who could grind another two tenths of a thousandth of an inch by giving the machine a kick! He had been working there – same machine, same spot, for thirty years he told me. He'd lost a finger, a thumb and had several scars from accidents with the grinding wheel. I was allowed to observe the safety committee at work and at one meeting they were discussing a recent accident where the machinist had grabbed hold of the grinding wheel when it was spinning at 2,000 rpm and had lost part of his hand. They wondered how an experienced operator could have done this. One committee member hesitantly ventured to suggest that the operator may have thought that the wheel was stationary and that this may have been an effect of the lighting. (This is a well-known effect, when a spinning wheel can indeed appear to be stationary when the rpm exactly matches the frequency of the lighting – i.e at 50 cycles, a wheel will appear to be stationary at 50, 100, 150... rpm – a similar effect that one observes in films when car wheels go backwards or appear to be stationary). The chairman, a large Yorkshireman, bellowed out that he would have none of that silly scientific nonsense here and let's just stick to facts. I was frustrated beyond measure, as it was clear to me that that was exactly what had happened but I was an observer and not allowed to say anything. The committee allowed the operator £10 compensation! Another chap, Jock, a lovely Scotsman, had walked down from Scotland in the Thirties with the Jarrow marchers to find work when they were all unemployed. He was a very interesting and intelligent chap and we remained friends all the time I worked there. Another machinist I worked with was a chain smoker. He rolled his own, quite expertly, with one hand. As one cigarette went out the next one was rolled and lit from the expiring one. They were poor-looking fags, rather limp and thin. His hands and face were lined and like leather, brown with nicotine stain and he had a constant cough. He would spit regularly into the machine lubricant trough. He put me off cigarette smoking instantly. I had been smoking a bit from about eighteen but from that day I never again smoked a cigarette.

For a while I worked as a fitter and setter, and I had to repair and keep running the hydraulic presses used for bearing assembly. The operators were all young women and were a bit rough around the edges – literally. I recall the bitten fingernails in particular. I was told that I would lose my trousers in a flash if I so much as passed by the department. I didn't, but the girls used terrible language and told the filthiest jokes I'd heard. They were on 'piece work' where they were paid on results, so when their machine broke down they would shout to me, "Derrick! Get your arse over here and fix this f... bleeder". I had a lot of laughs with them and heard about their rather humdrum lives in Chelmsford. I quite enjoyed this short spell (in fact, I enjoyed all of these different experiences).

In the design drawing office I had to do extremely long calculations and the only help we had was a hand- driven calculator! Later we acquired an electric one, but it took a long time to do the calculations nevertheless. At college we used slide rules and we became quite adept at using them. Hand-held electronic calculators were still a few years away. I decided around this time that sitting at a drawing board was not for me in the long term.

In the technical advice department we had to respond to customers' enquiries. Many requests were humdrum but others ranged from what wheel bearing to use in a four-wheel drive car to rocket-assisted plane launchers on an aircraft carrier. This last one I remember as the wheel taking the steel hawser from the rocket to the plane accelerated from zero to an extremely high speed in a few seconds, and the stresses on the bearings were enormous. The predicted life of the recommended bearing was rather short and they would have to be changed at set intervals. I had to calculate the stresses and make a choice of bearing type and size to suit the application. I sat at a bench, and wrote my response in pencil on cards which were vetted by a supervisor. He was a stickler for the prose, in the "we were proceeding in a northerly direction" style, which wasn't mine. Once we had agreed on the result, it was typed by the typing pool, so I missed out on the relationship with the customer and I never found out whether they bought the bearings.

I worked for a short while in the metrology department – this was an underground, vibration-proofed, dust-protected, temperature and humidity controlled room with highly sensitive measuring instruments of all kinds. Samples were routinely taken from the production lines and checked over. The apparently smooth bearing race surfaces, when magnified and measured to a few millions of an inch looked like the Swiss Alps! We also checked the competitors' bearings and logged their results. Hoffmann bearings were considered to be the Rolls-Royce of the bearing world but gradually the competitors were catching up. The Japanese bearings that we checked at first were laughable – dreadful quality. But over a breathtakingly short time the quality improved by leaps and bounds. After I had moved from this department I would pop back in from time to time and I saw that there was really cause for alarm. The Japanese products were on a trajectory that would see them overtake our products before long! And that is precisely what happened. I don't know what the management were doing but it certainly led to disaster of the first magnitude. What was the point of all this equipment and checking, if no action was taken when needed?

Between these industrial work periods I attended college in Enfield. The technical college soon became Enfield College of Technology (1962) and is now the Enfield campus of the Middlesex University. For the first year I was in digs in a small terraced house. For £5 a week I had bed, breakfast and evening meal. This left me with more than £3 for everything else. The landlady was, I discovered, a Jehovah's Witness and obviously thought that I was ripe for conversion. She started on me in a rather unsubtle way. When I responded by asking her what happened when Heaven was full (they believe that only one hundred and forty four thousand can get there), and perhaps it was full already, she was stuck for answers. So one evening on arriving back from college I was met by three serious-looking gentlemen in suits who, it appeared were senior members of the local JW church. I proceeded to discuss their beliefs which sounded totally preposterous of course. Again they had no answers to my questions, especially when we got onto the Creation. The world, in their eyes, is only a few thousand years old, so when I raised the matter of fossils and Darwinism their only response was "God moves in mysterious ways" to which I said that they'd need to do better than that. They left, clearly thinking that I would go to the devil.

It was a three-year course and for the next two years I shared flats. This was much more fun. Along the way I sold my scooter, bought a better one, then bought my first car for £180 cash, a Standard Eight. This was basic – an 803 cc engine, three gears, no heater, access to the boot only from inside... but a big improvement on a scooter for bad weather – and for getting around with the girls. One flat was in Green Lanes, north London and was above a tobacconist. It was frightfully cold that winter, with smogs, fog and much rain. One night strange noises in the road aroused me and I got up to witness the removal of the trolley bus overhead lines. The lines were cut from the poles with shears and went twanging into coils. The end of an era. I shared this flat with three others and it worked out well. One Sunday evening flatmate Andy came in late from a weekend away, proudly holding a sack. His girlfriend's mother had donated two chickens to our larder! They were complete with heads and looked as if they'd died of consumption. As the only one present with any experience of dealing with dead chickens it became my task to prepare them. I did my best and they supplemented our modest diet but there wasn't much meat on them. This was the time of the infamous London smogs and one evening we had to crawl along the A10 Great Cambridge Road at walking pace with someone walking in front of the car as the smog was so thick that I could only just see the front of the car bonnet. Abandoned cars littered the road and police used flares at junctions. Coal-burning fires were banned some years later.

At the junction of the A10 with the North Circular Road I witnessed an accident late one evening as I was filling the petrol tank of my scooter. Hearing a bang, I looked up to see a car flying through the air, upside down! It landed in the middle of the roundabout (now replaced with an underpass), on its roof. I told the garage cashier to call an ambulance and the police and rushed over the road to the scene, to find four people stuck in the car, with two young women in the back, screaming. There was a lot of blood and petrol was leaking out of the tank. Fearing a fire, I spoke to the driver through a broken window and got him to turn off the ignition. Fortunately, no-one was smoking! After a struggle I managed to open a rear door and got out the two women who were hysterical. Although there was masses of blood I couldn't establish any wounds, until I realised that one of them had torn off a piece of ear lobe. The young men in the front were pulled out and then the ambulance appeared and I left the professionals to it. I was the first on the scene again later at two similar accidents, both involving cars landing upside down!

I became involved with the Student Union and became the film club secretary before taking on the job of social secretary. Running the film club was easy, I was told – just pick a film and advertise it. Membership of the student union was voluntary so we never had much money and suddenly the film club became a potential source of revenue – and profit. By judicious advertising we gained good audiences and made money. I became ambitious and selected "*All Quiet on the Western Front*" for a showing. In the canteen I had a large poster "Watch this space". After a few days another appeared "It's coming". A few days later another said "Very soon... " and the final one "It's arriving – *All Quiet on the Western Front*". The tickets sold like the proverbial hot cakes. The film, however, failed to arrive by post. The postal service had gone on strike and parcels were piled up by the thousands in the sorting offices. I was frantic, and went down to Soho Square to the film company's office. "Sorry, chum," I was told, "we have only the one copy of that film". In desperation I brought back a W.C. Fields film. To a packed audience in a lecture theatre I had to front up and tell them the news. I was catcalled, booed and hissed off the stage. I offered to refund monies but very few took up the offer, even though W.C. Fields didn't go down at all well. We never did get to show the famous First World War movie.

As social secretary I organised dances at the college and hops at the sports pavilion at Winchmore Hill. This was opposite a teaching hospital so we had a source of girls to invite. One evening I dated a fantastic looker – she was Spanish but her English was poor and my Spanish non-existent so we didn't get on too well. We had some great college dances and again we made money as attendances were good. I took risks in inviting some of the big jazz bands of the time as I didn't know whether we would take sufficient money to pay their fees so I learned a little about risk taking and advertising. Eric Silk was a great success and then I got

Mick Mulligan with George Melly. They were a riot with George (with three women in tow) in top form. During the interval he produced a bottle of whiskey and proceeded to drink it, then took some crayons and quickly sketched up a cartoon on a wall backstage. It was still there when I left the college. Shortly after this gig he dropped out of live performing, becoming a writer and critic for some years before getting back into performing. He died in 2007, I was sorry to learn.

The Student Union president would meet the college principal regularly to discuss matters of concern. One matter that cropped up was car parking for students. This had not been a problem before as students didn't have cars but now we were beginning to acquire them, and as engineering students we would buy very old, cheap cars and get them working. We had a field to park them in but this rapidly became a quagmire in the winter. Our entreaties to do something fell on deaf ears so we devised a cunning plan. We drove into college early one morning and took all of the lecturers' and staff parking places, including the principal's! There was a rumpus, of course, but having the staff all walking in with mud up to their knees made the point beautifully! The field parking area was quickly covered with clinker obtained from the local power station. This type of problem-solving later featured in one of my training courses – "If the owners of the problem won't do anything about solving it, do something that makes them experience the problem, then they will solve it!"

Around this time I fell in love for the first time, and I experienced the heart-stopping thunderbolt. I had had a number of girl friends since leaving school (five of them named Carol or Carole), none of them serious. Carol number five was at the Cockfosters Teacher Training College and I just went weak at the knees as soon as I saw her at a college dance. I zoomed in on her like a guided missile and we spent the entire evening together. We got on famously for the best part of a year. She was in a student hostel at first and they locked the doors at night so we had to get used to climbing over the high brick wall surrounding the place. I went to her home at Surbiton and I took her home to meet my parents at Clacton. We were in love and I was walking on air, as they say. We spent some weekends exploring London, one day visiting St Paul's Cathedral. We climbed to the dome, experienced the Whispering Gallery, then climbed higher, eventually reaching the cross. We climbed into the cross and looked out onto London's vista through a grill at the centrepiece of the cross. I don't think that this is allowed these days. It was a very special day for us. Somehow things fell apart, we broke up and I was heartbroken.

In the meantime I was trying to apply myself to my studies. I wasn't a particularly good student and really needed to work hard to succeed. The workload was quite tough with long lab reports to write up on the practical work apart from everything else. It seemed like the art students had so much less work to do than us engineers. My worst subject was electrical technology. In fact, at the Swots Ball at the end of the course, the lecturer gave my wife-to-be the advice that she should never let me rewire the house! (I actually did rewire my house several years later!)

Every year during my teens I had suffered from a bout of tonsillitis. This always laid me low for a week or two and the last straw was getting a bout of it just before my finals. I was feeling wretched just as I should have been revising, and I just about recovered in time for the first of the exam papers.

My pass wasn't as good as I would have liked and I soon fronted up to my doctor and demanded that my tonsils and I parted company. He had been reluctant to recommend this previously but I insisted. There was a large waiting list for operations of this type, he told me, but if I could make out a good case he might get me into the military hospital at nearby Colchester. I had no trouble making out my case as I was absolutely fed up and no doubt I embellished on my agonies. I got in within a few days and found myself alongside soldiers, sailors and airmen who were mainly being treated for the results of exciting accidents, like falling out of planes, being accidentally blown up or being hacked at by a machete-wielding terrorist. I felt a bit of a fraud just having my tonsils removed, and would have been happier with at least a blood-stained bandage around my head. The tonsils that had caused me so much trouble were removed without drama. As I was recuperating I was informed that the matron was about to visit our ward. Beds were tidied, the floor was cleaned (even the sick and lame had to get out of bed to help clean) and everything was spick and span when the matron and her entourage arrived. She swept down the ward in full regalia like a battleship under full steam with an attendant escort of destroyers. She paused here and there, her eyes missing nothing, while the patients stood to attention by their beds, some on crutches, eyes front. The escorts looked worried and remained silent. She arrived at my bed, paused and turned to face me. Being an outsider to all of this, I was sitting up in bed in my military-issue pyjamas reading my James Bond book, munching grapes. The matron was a large woman, very large. She was almost as wide as my bed, her immaculate uniform straining to contain her huge bosom. Embroidered epaulettes sat on her shoulders and some formidable headgear gave me the impression that I was looking at someone who wielded power by the bucketload. She was probably at least a major, if not a general. I gave her what I thought was a winning smile. "Good morning, matron", I said. There was a silence for some few seconds as she looked at me with cold eyes. Then she opened her mouth. Her voice could have carried all around Colchester without amplification. "Get out of that bed!" she bellowed, "and stand to attention when I approach you". "Er, I'm not in the forces", I said, "I'm a civilian". If looks could kill, I would have been in the mortuary. "I don't care what you are," she responded with enough force to almost shatter my ear drums, "get out and respect my authority by standing to attention!" I meekly obeyed this instruction, thinking that I could be shot at dawn by this creature. If she had been born a few years earlier she would have cowed Hitler's armies into submission all by herself.

I recovered from this trauma and made progress such that I was told I could leave hospital the next day. The nurses attending the patients were from the Queen Alexandra's Nursing Corps and I had been chatting to one of them, a pretty young thing who had caught my eye. When she came to make my bed I got out and helped her, giving her my best line of chat. Returning to my bed when she had left, the sergeant in the adjacent bed asked how I had got on. "OK," I said, feeling rather pleased with myself, "I've got a date for next Saturday." "Well," he said, "I think that you really impressed her." "Yes?" I said "why do you think that?" "Because," he said, "for a start, your arse is hanging out of your pyjamas!" My pyjamas had split along the rear seam from the waistband to the crotch and I had been bending over the bed in my full glory!

At the end of the course I returned to Hoffmann with my Higher National Diploma (HND) in mechanical engineering. However I'd been invited to apply for a fourth year at college – to take a Diploma in Management Studies. I asked my employer if they would sponsor me – and to my astonishment they agreed. There were only three of us in my year sponsored by Hoffmann and I was the only one to take up this fourth year. I enjoyed this year the most – management subjects including applied economics, industrial law, management of men, (women didn't seem to count), applied statistics, engineering design – and I gained a diploma with distinctions. I enjoyed the studies much more than the previous years, and with hindsight this was a reflection of my real interests. Some of this material was quite valuable to me later on.

It was during this final year that I had my introduction to the emerging world of computers. We had to do a final year project and I chose to research the phenomena known as the 'hydraulic jump'. This is a rather strange event that occurs when water travelling at high speed slows down. From a low-level stream at high speed it becomes a slower-moving but deeper flow. At the juncture where the change in depth occurs you get a sort of backwards waterfall. It's rather strange to observe, and my task was to verify that the formulae put forward that underpinned this were sound. The apparatus that I used to check it was formidable – a large glass channel with a weir apparatus, (a glass plate that blocked the channel except for a gap at the bottom), weighing machines, pumps and a large quantity of piping. The whole set-up took up most of the laboratory with many gallons of water being pumped through per minute. Over several months I took readings and made adjustments to check out the theory. However, the formulae involved were most complicated and with only a slide rule it took several hours to run just one set of my results through the formulae, and I had about thirty sets to process. Then out of the blue we were suddenly given the opportunity to attend a two-day course on computers. A presenter from what was then ICT gave us a course on programming, using a very basic computer language. The course was focused on scientific and mathematical applications and we were invited to come up with an application to program, and we would then be able to run it on a computer. And did I have just the application! I was able to write the program for my formulae, then I took it into the ICT building in Newman Street in London where it was punched onto paper tape. My test results were also punched onto tape, the program was compiled, it seemed to be OK, so I ran it. A minute or so later the printer burst into life and there were all my results! I was amazed, and that of course was the best introduction to computers that I could have had. The computer was the size of a small house and memory was tiny, just 1K, I believe, but this was 1963.

I had an embarrassing incident during this time with the project. I had set up the equipment for a new set of measurements, but before I could take them I had to let the system settle down, and this took at least twenty minutes. I wandered off into the canteen for a coffee. However, I hadn't balanced the output flow to match the input flow, and slowly, slowly, the output reservoir filled up... and overflowed. I was taking a leisurely break and when I returned water was running out under the laboratory doors. Opening the doors released the flood, as the lab was now several inches under water. I wasn't the most popular student that day.

That year I rented a terraced cottage at Winchmore Hill with three friends. It was a great little place, a basic "two up, two down" which had been extended. The front door opened straight into the sitting room that had a pot-belly stove and was very cosy with a fire going. The whole place had been nicely done up and we appreciated it. The cosy sitting room was the "seduction room" and when we had a suitable date we would book it so the others had to make themselves scarce on these occasions. We put a red light outside which I'm not sure that the neighbours appreciated but on the whole I think that we weren't too bad to live alongside – although we had parties they weren't too rowdy. After one party someone left a newspaper parcel in the bedroom that I shared, and it remained there for some time, as each of us thought that it belonged to the other. A bad smell became apparent, and again we both thought that it was the other's socks, so said nothing. Eventually we deduced that the parcel belonged to neither of us, and that it was the source of the now very strong smell. The parcel contained some old bones, and they were decidedly off!

That year we had heavy snowfalls and very cold weather. One of my cottage-sharers, Andy, had an old Austin Seven, and he had left it on a ramp on the Thames at a rowing club. The tide came in and the car was flooded and almost floated away. He managed to save it and with much care he got it going. However, water in the petrol kept causing it to stop and he developed a neat routine in emptying the carburettor and restarting it. Another problem was water in the gearbox. After a cold night the car refused to start and he couldn't turn it over with the crank handle. The gearbox was frozen solid! I was running my Standard Eight and we used the cars to run back and forth to college. We had to dig them out of the snow several times that winter.

CHAPTER 4
Early Career and Marriage

It was during my final year at college that I met the woman that I would marry. Jennifer was the assistant librarian in the college library and I met her while looking for the books to help me with my project. We starting dating and it became serious. By the end of my final year we were an item. On returning to Hoffmann in Chelmsford it was clear that things were about to change. I had been living at home while working in Chelmsford (home was now Jaywick, where my parents had moved to in 1959, having sold the guest house) and now I was frequently travelling to Barnet where Jennifer lived. I drew a circle on the map based on Barnet, assuming that we would live somewhere around there, and identified all possible engineering companies within a radius of twenty miles. I wrote to a number, offering my services. I had an immediate job offer of a design draughtsman from Scania, the truck manufacturer, without even wanting to interview me! I wasn't very impressed by this, and declined on principle. However, I had several offers for interviews, including one with Handley-Page aircraft at Boreham Wood. I went for an interview, which seemed to go reasonably well and they made me an offer of a job in the wind tunnel test department. Then I looked around at the bored-looking chaps sitting at their drawing boards and after several minutes thought, I declined! I finally took a job with the Electrical Apparatus Company (EAC) at St Albans. I made my farewells to my mates at Chelmsford and thanked Hoffmann for my sponsorship. I'm unsure now if I felt bad at this – after all, they hadn't had much useful work out of me and they had paid for me to study for four years. But there was no contract and this was a risk that they were fully aware of, and was fairly common then with the larger companies. I followed the fate of Hoffmann for some years – they gradually declined, in common with most of the British manufacturing companies, and then merged with two of their competitors to become Ransome Hoffmann Pollard in 1969. They are now owned by NSK but the RHP brand remains. The Chelmsford site is now a housing estate.

My new employer was a private company, employing about one thousand five hundred people that made switchgear. I was to be a member of the works management department which consisted of three people – the works manager, the assistant works manager and the secretary. The two managers were nearing retirement and the unspoken assumption was that I was to be groomed for one of these jobs. As a temporary measure I again took a house share with some of my college mates in a large, rather grand property in Winchmore Hill, N 21. I was the front man of the group and the rather frail old lady that I met when I went to check it out was sweet and totally unaware of the risk that she was taking in letting her house, with its grand piano, fine furnishings and delicate knick knacks to a group of young men. I managed to persuade her to put the grand piano into storage, as I visualised cigarette burns and beer glass stains or worse on its highly polished surfaces. I slept on a camp bed in the sitting room for some months while Jennifer and I sorted out our domestic arrangements. Although the house let was ostensibly for four, in practice there were at least six living there, sometimes eight or nine. My housemates and I shared the cooking arrangements and took turns at preparing the evening meal. We didn't have a sufficiently-large saucepan so we got hold of a wash bucket, the oval sort that has a strainer to squeeze out the mop. In this we prepared stews and curries. Any remainder would be left in the pot and added to the next day for the next meal. Sometimes weeks would pass in this way, but no-one appeared to succumb to any tummy bugs.

Jennifer and I were determined to buy a house or flat from the beginning – renting seemed to be a waste of money and a situation that it would be difficult to change. However, the price of north London property and the lack of any sizeable funds made this impossible. I had just £200 saved in a post office account and Jennifer had less. So we widened the scope and looked at Luton. It was not a very inviting place to live but it had an attraction – it was on the recently-opened M1 motorway and housing was much cheaper. For £2,200 pounds (!) we could buy a newish small, very basic two-bedroom semi-detached house. I went to Nan in Clacton for a loan, for she was the only one with any sort of cash to spare. She immediately produced £200 in notes from somewhere and we agreed on repayment with interest when I could manage it. It took several years, but I repaid it with the interest. We bought the house, and I performed the paperwork. At least, I was the front man. I wasn't legally qualified of course, but my future brother-in-law was a solicitor-in-training, and although he was unable to officially take on the job of conveyancing, he was able to tell me what to do. When I said that I was doing my own conveyancing everyone would tell me that I had to have a solicitor. This was untrue, it was perfectly legal to do it yourself but few knew this and fewer approved of it. It all went OK and I attended the little ceremony where money and paperwork were exchanged – and the house was ours! We got married at Barnet Registry Office and had a reception at a local hotel, on March 14, 1964. For our honeymoon we chose a week touring around the Cotswolds in my Standard Eight. The weather was terrible but we went to Stow-on-the-Wold and the other beauty spots and stayed in small inns. Jennifer became sick within a few days so we had to cut short the honeymoon and return to our small house in Luton. It was rather cheerless at first, as the house, although quite new, had been occupied by a dysfunctional family and was rather knocked about. We had lots of "overdue" notices and debt collectors knocking on the door for some months, chasing the previous owner, who had fled to an unknown abode. Our furniture was basic. We decided not to buy anything on hire purchase, except for the gas stove, so we had only a bed, table and chairs and second-hand settee. Our first meal was eaten off the dustbin lid as the table hadn't arrived! I began to try out my practical skills and worked on the plumbing to bury water pipes in the wall and I installed extra power points. I installed a clock point for the kitchen clock and it worked – until I switched on the kitchen light, when it stopped! My wallpapering skills were tested when I papered the whole place including the hall and staircase. I was completing the latter when I did a Laurel and Hardy act – I stepped off the ladder into a bucket of paste! So we settled in, Jennifer took a job as the local librarian at Sundon Park library which was just a short walk away, and I drove down the M1 to St Albans to my new job.

At EAC I did the rounds of departments to get to know the place, working in each for a month or two – time study, production engineering, mechanical/electrical design.

The works manager and the production director who had hired me gave me projects to work on as I went. I applied for membership of the Institute of Mechanical Engineers but found that I was still short of the required practical experience – they were sticklers on this – so I worked on production for a few months, wiring up control boxes. The wires were thick and heavy and the wiring had to be neat and tidy. We wore gloves to keep the wires clean-looking – the white wires especially would show fingermarks. I became adept with the wiring tools and I began to learn about the contactors and switching equipment that the company made. It made control equipment to switch heavy currents for power stations and industry generally and there was plenty of work. One day the works manager gave me a problem to solve. We had a contract with a Defence department and we were behind schedule. Apparently we had a production quality issue, which caused units to fail in test. Unless we could fix this soon, we would default on the contract, and be heavily penalised. I looked at the production problem which was caused by worn-out tooling on an operation where we modified a bought-in component. The new press tools were being produced but would take some weeks, and the delay was unacceptable. I investigated whether we could reduce the tooling manufacture time, but could make no useful improvement. I designed several alternative components that I thought would do the job, but they were all rejected by the chief designer. Then I spoke to the manufacturer who supplied the basic component. "Do you make variations on this?" I asked. "Sure do," came the reply. "Could you make one like this?" I asked. I got my reply within a few days. Not only could they do it, they could supply a thousand or two within a week – and at the same price as the basic component! We scrapped the tools, the modifying operation and the three full-time jobs that it entailed. That was my first experience of "but we've always done it like that" and the value of asking "Why do we do this? and "What else could we do?"

Another project involved a woman who collected statistics from the factory floor and spent a week analysing them. The results were written onto a large sheet, (a sort of spreadsheet in fact) the size of a drawing board. When I asked who used this result, I was told that it was only looked at by the works manager. He only used a couple of the numbers of the final result, I found. I was able to simplify the system so that the whole job could be done in a half-day. Another case of "I've always done it this way, so why change!" These examples were two of a number of projects where I used a technique that I had learned while studying work study. It was based on the "what, why, where, when, how" questions and was so effective that I used it later in my computer career with great success. I eventually included it in the training courses as well.

I was next asked to look at production issues at a subsidiary company, just a mile or two away. They couldn't keep up with their schedules at all, as they were always running out of stock of the components. Any product has a hierarchy of components, some of them manufactured in-house, many of them purchased. When a customer places an order for the product a manufacturer must be able to place purchase orders with their suppliers and to plan the production and assembly of the in-house items, if they are not in stock. The production scheduler was rather reluctant to discuss things with me, but he really had no choice as both he and I reported to the production director. He saw the ground disappearing from under him. He had good reason for concern – he was inept at his job, and he had no good system to keep track of components needed. I built a manual (paper) system (known as 'bill of materials'), rather like a spreadsheet, so that he could forecast and obtain the needed components in time to meet production. Something similar was needed at the main plant as well. I then began to realise that any manual systems had weaknesses – namely the mortals that had to operate them – and that the computer that had solved my problem as a student could do the same for this company. I began some elementary research, and I made a recommendation that the company should invest in a computer system. This was still early days in the field of computing but there were computer systems and software coming into use in industry. My knowledge, of course, was thin to say the least, and my experience was zero. I didn't know much about presenting a case as large as this one and my recommendations were politely ignored. However, I had been intrigued by what little I had learnt and saw the potential for computing to make a huge contribution to the sorts of problems that I had seen. The more I learned the more I was interested. Perhaps a computer career could be more suited to my temperament and more rewarding? Engineers were not particularly well paid, it seemed to me, and Britain didn't seem to value them very highly. Manufacturing in the UK was on a downhill drift and although I had slipped into production engineering rather than mechanical engineering, and I was finding it more interesting, I seemed to be attracted by systems – the organisation of people and the processes of making things work.

I applied for a few jobs with the computer manufacturers but soon realised that they were focused on the hardware. That wasn't what I wanted. Then I answered an advertisement that described me exactly – and landed a job with Marconi Instruments, very conveniently situated at St Albans. At the interview I was asked if I knew anything about purchasing and the creditors' ledger. I said that I did. Well, I knew a tiny amount about purchasing, but I had no idea what a creditors' ledger was or did. I was soon to find out as my first project was to design, develop and implement the creditors' ledger system! I joined a small systems section of the data processing department in early 1966. This turned out to be quite fun as I and one other engineering graduate formed the nucleus of the systems analysis section. I did a couple of basic courses but as computers were still new it was the blind leading the blind. The company had some punched card equipment, ancient IBM machines that processed 80-column punch cards by the thousand. Our job was to design systems to firstly take over the punch card accounting systems, then to further develop systems that were really needed by the company – like production control. I established that the creditors' ledger existed to keep accounts of the company's debts to their suppliers, and to pay them. It wasn't too difficult to design a system to do this, and to explain this to the relevant people. I had to learn some accounting jargon and I soon learnt not to be embarrassed by my lack of knowledge. As soon as someone mentioned some jargon or acronym that was unfamiliar to me I would stop them immediately and ask "What does that stand for?", "What does that mean?" or 'How does that work?" The most important question of all though was preceded by "why?" I found again that some processes were performed because they had always been performed. When I had written the system specifications the programs were designed and written by our programmers. Then we had to test the programs. A system consists of thousands of program instructions so there are going to be errors (these days systems contain millions of instructions). As our computer – a System 4 from English Electric Leo Marconi – had not yet been delivered, we arranged to use a similar machine running at Stewarts and Lloyds steelworks in Glasgow for the testing. I then worked in Glasgow for eight weeks. Two or three of us would fly up on Monday and return on Friday. We had to sleep during the day, as we could only use the machine at night. My system proved to be reasonably free of bugs and I had it working within my time frame.

On one of the return flights to Heathrow in a four-engine turbo-prop I noticed with some alarm that one of the propellers was just going around slowly. The pilot came on over the speakers to inform us that he had shut down one engine as it was overheating. "There's nothing to worry about," he said, "we can fly happily on three engines but it will take longer" and he gave us a new time of arrival. Soon afterwards he came on again to say the same thing about another engine, giving us another estimated time of arrival at Heathrow. The fellow alongside turned to me. "I hope that the other two don't have to be turned off as well," he said "or we'll be up here all bloody night!" When we did arrive at Heathrow there were fire engines and ambulances lined up alongside the runway, but we made a perfect landing.

Glasgow was still a bit rough then and we'd been told to be careful. One night we approached the computer building which was locked up. We had to attract the attention of those inside by standing on a brick ledge and banging on a window of the computer room. One of my colleagues was ahead of us and he put down his briefcase and climbed up. By the time he'd got down – there was no briefcase. The road seemed deserted apart from us, and we had just appeared from around the corner. We were mystified and informed the locals inside. "Your briefcase will be in the pub opposite," he was told. "I'll go and get it then," he said. He was politely told that if he valued his skin he would do no such thing. He didn't get the briefcase back.

Back in St Albans, with the new computer installed, I ran further tests and did pilot runs before implementation. On one occasion the chief accountant arrived just as I was test printing some cheques. "How do we put a new supplier onto the system?" he asked me. "Easy," I said. "You just make out a document like this" and I took out a form that I'd designed. "Show me!" he said. "Just fill it in – put your name in." So I made out a form for the Derrick Brown Consulting Company, with address and so on. "Now put it on the system," he said, so I did, keying in the data. "Now an invoice arrives," I said, and I inputted an imaginary invoice. Then a 'goods received' note was input. "Now we'll pay the invoice," I said, as the computer system matched the invoice and passed it for payment. Out popped the cheque a few moments later, on the rather fine stationery that I'd designed. "Looks good," he said. "I suppose that you could go and cash this now?" and he raised an eyebrow. I nodded, then slowly realised the full meaning of this little enactment. I had designed a working system but anyone with a modicum of knowledge could use it to defraud the company! I went away to install the necessary security measures into the system. Computer fraud became big business and payroll systems and cheque-paying systems like mine were easy targets initially. We all had a lot to learn. Frauds were (and still are) mainly committed by those insiders who know how to get around the internal checks and security measures. It's actually quite difficult to make systems foolproof without adding layers of bureaucracy which is just what we were designing out. My system went live over the space of a few days, and began to do its work. I was quite thrilled to see the results of my months of work take effect. There were some queries from the accounting staff but the dust soon settled and I was onto the next project.

By this time Jennifer and I had moved from Luton to Oakwood Drive, St Albans, just a short walk from the Marconi Instruments factory. So close, in fact that I could walk home for lunch! We sold the Luton house after two years for £3,800, making a profit of £1,600! This was better than renting! The new house was a three-bedroom semi, with a good-size garden that backed onto playing fields and had a shared drive. A big improvement was a garage where I built a workbench for the home improvement projects that I soon got involved with. With my improved salary we could now afford a bigger mortgage. Jennifer was also earning better money as she had changed jobs and now worked for IBM in the city. This didn't last for long as she didn't like it much and she took a job as a mobile librarian, with Hertfordshire mobile libraries.

St Albans was a very pleasant town to live in, with the remains of the Roman city and its amphitheatre, the cathedral, river and parkland. It boasted the oldest inn in the country (one of several with this claim) with an octagonal structure. There were some good pubs in the city and it was after a meal at one of these on a Saturday evening that Jennifer and I became involved in a nasty incident. A crowd had gathered around a brawl, and we could see a person on the ground and another putting in the boot. The 'booter' was a large aggressive bloke, a six-footer and burly. He looked to be simple, without any intelligence at all. He had a companion, a small fellow, who was clearly the 'brains' of the duo. I was reminded of Lenny and George, in 'Of Mice and Men" except that this Lenny was a dangerous gorilla (sorry gorillas, I know that you are generally gentle and you look intelligent, let's say that this fellow was a slightly smaller version of King Kong). The chap on the ground was trying to protect himself as the gorilla kicked him without mercy. No-one in the crowd did anything. Jennifer whispered that I should do something. I didn't feel at all like taking on this fellow, who looked as if he could have me for breakfast. However, I had no option when Jennifer pushed me forward with some force and I found myself in the small clearing, with the body at my feet and the gorilla facing me. He looked at me in some surprise, and stopped the kicking as his brain tried to accommodate this turn of events. I, meanwhile, thinking that my last moment had come, could only think of shouting "Leave him alone!" very loudly and waving my arms about. The gorilla frowned, grunted and began to move towards me. Fortunately for me his keeper came forward, took his arm and led him away. I heaved a sigh of relief and turned my attention to the victim. Despite being quite battered and bleeding around his head, he was quite definite that he didn't want either an ambulance or the police, and after several minutes he got up and staggered off into the night. The crowd had melted away as soon as the excitement ended. Apparently this sort of thing was not uncommon in St Albans on a Saturday night.

Soon after moving there I had to use a company car. It was a large old Ford, with automatic transmission. I'd never driven an automatic before and this took a little getting used to. As I swung the car into the drive one evening I realised that the turn was not sharp enough, and I was going to hit my neighbour's garden wall. I stopped in time but as I fumbled with the gear change the car inched forward... the front bumper gently nudged the garden wall – and the entire wall slowly collapsed, crushing my neighbour's prized roses! He was not pleased, and I had to pay for a new wall.

Getting into home improvements I decided to remove the unused chimney breast in the kitchen thus making room for the central heating system that we planned. The chimney had housed a boiler, now gone, and became an outside flue at ceiling height. One Saturday morning with an empty house I started knocking out the bricks. I started at the bottom, figuring that they would just fall out without too much hammer work. And so they did. I was on the step ladder, with only three rows of bricks to go, as I cut through a brick that was half in the adjoining wall. With dismay I saw the three rows move – they were now unsupported, as I had just cut through the last means of keeping them up. I leant across the top of the ladder and held up the rows of bricks. Perspiration dripped from me as I pondered the problem. I needed some timber to support the bricks but there was no-one to call, no-one around. The floor around me was covered with bricks and dust. I did the only thing left – I jumped from the step ladder, moving as far away from the resulting mayhem as I could. The bricks fell – and so did the builder's rubbish that was piled between the ceiling and the floorboards above. I was lucky not to break any bones, but I was a bit battered and covered with dirt and dust. A second problem now appeared – three ceiling joists had been supported by the chimney breast and were now unsupported at all – just hanging in space held only by dint of being nailed to the floorboards above. As if that wasn't enough a third problem became apparent. I realised that directly above the unsupported joists was the bathroom – and the cast-iron bath! At least it wasn't full of water! I hurriedly put some timber supports in place then rushed out to buy suitable timber and the necessary things to make and attach extensions to the joists. By the time that Jennifer returned from her shopping trip I was just completing the timberwork but I still had a large hole in the ceiling. The kitchen was none too clean either.

"How did it go?" she asked. "Pretty well," I answered, "there were a couple of problems but nothing I couldn't fix!" Sometime later I was talking to the next door neighbour of the adjoining house. "I'm so pleased that you didn't take out the chimney breast," he said. "My wife and I were discussing this and thought that it could have been dangerous and perhaps unsettled both houses." I coughed and said, "Well, actually... " They moved a short time later.

I joined the company's amateur dramatics society and appeared in "Watch it Sailor!" This had been a film "Sailor Beware" with Peggy Mount playing the mother-in-law to be and Shirley Eaton playing the would-be bride. I played the sailor, Albert, who is engaged to be married to Shirley, the daughter. We had a lovely lady who played the Peggy Mount character and a good supporting cast. It was a romp and we actors got more fun out of the performances than the audience. At the dress rehearsal we had an audience of old age pensioners who had a free evening and were great – they laughed at everything. I went wrong and missed out a couple of lines but otherwise it went well. It ran for four performances and on the last night we decided at short notice to have a party. As our house was close I volunteered the location. We got in a barrel of beer and some other drinks, and everyone of course brought more. It was a great party. The personnel manager of the company was in the play and at the party. He was a rather large fellow and normally a bit reserved. However at the party he let his hair down and had plenty to drink. When he left he missed the front garden path, walked across the garden and straight into the hedge. Instead of backing out he fought his way through onto the pavement. Turning back to me standing at the front door, he waved, "lolly party," he said and went off down the road looking a bit the worse for wear.

We both liked dogs and now seemed like a good time to get one. A basset hound named Calamity Jane joined the household and was quite a character. She got on well with the cat and they would cuddle up together. Basset hounds are delightful, very good with babies and children, and provide endless amusement. Calamity Jane's coat seemed to be several sizes too big for her, and was full of wrinkles. It seemed to slide around with few attachments, if any, to the body inside. After becoming used to this addition, we decided to extend the household some more. Jennifer became pregnant very quickly. Lucy was born on August 4th, 1967 at Welwyn Garden City Hospital. The delivery became a problem, I was ushered out of the delivery room and Lucy was delivered with forceps, while I slept on a chair in the corridor. What a thrill it was as I saw her little wrinkled face for the first time and held her hand. I collected Jennifer and Lucy in our little mini and I'll never forget that moment when we arrived home and realised that we had this new responsibility and life would never be the same again. I decorated the nursery and built a patio with a sandbox for her. We had little knowledge of babies so had a lot to learn. Lucy was a delight, of course, and it was fascinating and a wonder to watch her as she observed us and the world, responding to us more as every day passed. The forceps delivery resulted in a red swelling under her left eye and this slowly grew and became quite prominent. One day Jennifer was very upset by an overheard conversation between two women who looked at Lucy and one murmured "the sins of the father...". We'd been told that nothing could be done until Lucy was older, but this was the catalyst to push for some action. The Great Ormond Street Children's Hospital gave us an amazing service, whereby the mark was injected a number of times over a period of some months to the point where it completely disappeared. Jennifer gave up working to be a full-time mum as was the custom then but she was never entirely happy about this and it later became a serious issue between us. Baby care facilities and nurseries were not well-developed at that time.

CHAPTER 5
Sainsbury

With our new responsibilities and only one salary it was important for me to ensure that my career was going well. After three years at Marconi I was looking for a new job. I found one at J.Sainsbury, the supermarket company. When I told my mother she felt that this was something that she could understand, and for some time I think that she thought I worked on the bacon counter. I joined the London head office computer department as a systems analyst. Sainsbury's occupied several buildings at the southern end of Blackfriars bridge and I worked in the main building, Stamford House. The building is no longer there. This turned out to be a good move (in fact they all have!) and I worked there for eight years from 1969 to 1976. The year I joined was the centenary year of the company. I landed on my feet, being given the job of replacing some old distribution systems. These ran on a couple of really ancient computers, one was a punch card machine, an IBM 1440, and the other was an EMIDEC. This machine was as large as a room, with mercury-delay lines, and just two K (two thousand bytes) of memory! We have gigabytes these days and more in a watch than that computer had. Eighty column punch cards had been commonplace, in fact they were often the sole means of inputting data and programs to a computer. By now they were slowly disappearing, being replaced by keyboards and other devices. The EMIDEC hardware was built on moving racks on rails so the maintenance men could work on it. They were constantly in attendance – it was always breaking down. The computer operators sat at a console like that of a modern-day airliner. It looked most impressive. Soon after I arrived we managed to get hold of an identical computer – the only other in existence – from London Transport, thus doubling our capacity. The computers ran the entire distribution systems, running for twenty four hours each day, except for the weekends. We were soon to be using the weekends as well. The files of data were held on one-inch wide magnetic tapes – there was no random access available, so data processing was all done serially and large sort programmes were required to get the data into the correct sequence. Soon after the new ICL computers arrived we had random access available, although the discs were huge platters and were held in cabinets the size of juke boxes. By current standards access was very slow and the storage capability was tiny.

I specified a distribution system that would take over from the old systems – there were three or four of them, all old and with many limitations – and it would cater for a new distribution depot that was being planned. I had to liaise with the new depot management, as they were planning some innovative ways of assembling goods. We also planned to transmit data to the depots and print the assembly documents there. They were presently printed at Blackfriars and taken by taxi to the depots! The new system took shape but to get it installed in time to meet the new depot opening meant that I had to work long hours with my team. The first ICL computer, a 1903, was delivered. During the testing of our new programmes it seemed to be running very slowly, and we spoke to ICL about it but they seemed to do nothing. Then we wrote a special test programme that did nothing except read in a record, add one to a record count, and then copy out the record to an output file. We read a few thousand records and recorded the results. This extremely simple programme used a large proportion – it was something like fifty per cent – of the central processor! ICL was incredulous but provided with the proof they went away. They quickly replaced the 1903 with a more powerful 1904 and quietly removed the 1903 from the market. These computers were large – taking up a number of cabinets the size of telephone kiosks, along with card readers and line printers. An air-conditioned room with raised flooring was required and the cost by today's standards, was astronomical.

We installed my system on time and it all worked. The new depot quickly got to grips with it and pronounced it "good". We then installed it across the other depots. We did however, experience just one serious problem after the system had gone live. One day the document printing went wrong – many were duplicated and we couldn't find any reason for this. Now my system was 'mission critical' – if it failed then the whole distribution system stopped, no goods could be moved, the warehousemen had no work – and the supermarkets would have no replacement stock. There was nothing more important than keeping full stocks on the supermarket shelves. The computer manager had received the phone call from our director, who had received a call from the retail director, who had received a call from senior branch management... by the time the message got to me I was informed that all depots were in trouble, that the branches were all out of stock but huge quantities of strawberry jam were being delivered! My manager had clearly accepted all of this without question and he looked as if he could see his job and retirement pension under the axe. In fact only one depot was involved, and one branch. I took over a meeting room with my team to examine the issue. The computer manager paced up and down the corridor outside like an expectant father and kept popping in to ask how we were progressing. I eventually went out to him and asked him to desist. I said that I'd give him a report at thirty minute intervals. That was less stressful for the team. We found the unusual error in the code and fixed it. We never had any other major issues with the system. This sort of investigation was and is quite common in software development and the skills involved are analytical, of course, like a detective's. I was often amazed at how many otherwise intelligent people would jump to a conclusion – often the first thing that came to mind – in these situations. "Well, it must be this," they would say, on the flimsiest of evidence and then waste time and effort pursuing the wrong solution. I observe this today in many commonplace situations. Problem analysis skills seem to be in short supply. What is that saying? "To every complex problem there is a simple solution – which is wrong." I included problem analysis and solution design in my training courses in later years but these were difficult to sell. Managers seemed to think that it was just common sense. Common, it was not!

I'd designed the assembly document that would eliminate a second document that was used when products weren't available, and this would save work for the assembly men and this was introduced as the next phase. This worked well and the new system was introduced into the other four depots. As the system was a major change for the assembly men I was asked to go to a depot to explain things. I duly turned up at a large depot at Buntingford in Hertfordshire to address a meeting of the foremen, who in their turn were to explain the issues to their respective shifts. When I arrived I was taken into a small room. "There's been a change of plan," the manager explained, looking a bit embarrassed. "The foremen have decided that they would rather you explained things to the men directly". I was a bit nonplussed at this. "When?" I said, "How many... " "Now," he replied, "we have two hundred of them sitting right outside." Pulling aside the sliding screen we were on a stage facing two hundred men in overalls, sitting there with folded arms, waiting for me to speak. I almost wet my pants. I gulped and asked for a drink of water. I had put together a small talk, complete with a few flipcharts, but it wasn't designed for an audience of two hundred! I took a few minutes to set up, aware that all eyes were on me. "Just talk among yourselves," I said as I thought quickly. I got going and got to the difficult part – they would have to change their procedures, and generally people are resistant to change. Before I had got far into it a large chap at the front stood up. "Why exactly are we changing things?" he said. "Why should we do this?" I gulped again. "What a good question," I said, "thank you for that." I realised then that I had missed out the vital part – the reason behind the change. Once people realise that there's a good reason for a proposal and can see some benefit, then they may consider it. I had to repeat this presentation to the other shifts, and at the other depots, but of course I was better prepared and it went down a lot better.

The next stage was to introduce the data transmission to connect the depots. I had recommended that we should use off-line equipment that wasn't directly connected to the computer. The computer would produce magnetic tapes that would then be read by tape readers attached to the transmission kit. The reason for this was that the ICL 1900 range computers that we were using weren't all that powerful and ICL had not really developed their data transmission techniques. We were going to use private 2400 baud (bits per second) transmission lines – very slow by current standards but then this was state-of-the-art. The equipment proposed had two unique properties that made the data transmission faster than any equivalent system, so I was keen to use this particular type, made by Standard Telephones and Cables. There was a strong feeling against this proposal from some in the department, who were emotionally attached to ICL (and who, I believe were sometimes wined and dined by them) as well as the ICL sales people who tried several tricks to torpedo my proposal. I put my proposal forward with feeling, as I knew that transmission speed was of the essence. We had thousands of documents to transmit and they had to arrive as soon as they were available. Hundreds of assembly workers would be waiting for them. The arguments ranged back and forth, the major argument against being that it seemed like a backward step to use off-line equipment. Eventually I had to argue it out in front of our director, who came out on my side. We installed the kit in all the depots and it worked well for many years. I found out that much later, in the Eighties or Nineties, when Sainsbury's had switched to IBM (and later experienced some disastrous computer system failures) that the staff were confounded when they transferred my old system to the new IBM equipment. The new system transmitted the data much more slowly! They couldn't work this out, but by then I was possibly the only person who would have known and remembered the two characteristics of the STC kit[1] that made it faster whatever line speed was used! As it was non-standard no contemporary kit would have had the same characteristics. It's an unfortunate fact that often the best ideas are swamped by lower quality products that are however, marketed more expertly. IBM computers were never thought to be the best in the market but they pushed all competition aside by their excellent marketing and a highly professional and ruthless sales organisation.

Although no-one had actually told me, I had become the de-facto project manager of this project that had grown so that we had a team of three systems analysts and a number of programmers. I found that my engineering background gave me an advantage over most of my colleagues – the engineering disciplines of development and design, coupled with testing and implementation processes all helped me. And the people aspects helped. I had become used to investigations, asking questions – and working with people at the coal face. Many of my colleagues fought shy of getting

---

[1] STC used a seven-bit byte against the standard nine-bit byte, therefore every item of data took seven ninths of the bits used in standard format. They also used a channel for constant verification of data transmitted, whereas the usual method of transmission used the 'ack/nack system (acknowledge/negative acknowledge), whereby the modems were turned around from 'send' to 'receive' and vice-versa to verify each block of data transmitted. While this was happening there was nothing being transmitted. These two characteristics gave STC an unbeatable lead over their competitors.

out to talk with the people actually doing the job (the so-called 'ivory tower' syndrome). Some of them never left the office. I had no such reserve and I became known at both management and worker levels in the depots and retail parts of the business. Consequently I was able to pick up the phone and get the first-hand response to any question that arose during the system development. I was particularly interested in the 'user interface' – how people interacted with the computer systems. I always paid great attention to this aspect, and tried to ensure that they understood the issues, and that the interface was designed so that mistakes and frustrations were minimised.

My first systems manager was an ex-army officer who had brought with him another old soldier as his deputy. The pair of them were quite hopeless and totally out of their depth. They understood punch cards but got lost after that! We (the staff) had been rumbling about their ineptitude for some time, and this filtered through to the director. One Monday we found their desks cleared – they were out, without any word to the staff, now about eighty strong. A colleague, John, a lovely chap was promoted to manager, and the first thing he did was to ask me if I felt miffed that I hadn't been chosen for the position, and did I mind working for him. I wasn't and I didn't, for the simple reason that I hadn't considered myself eligible for the position (I was still a bit short on confidence) and John, I thought, was a good choice. He was older than I by a few years and he had bags of common sense. He and I saw eye to eye and I was appointed deputy systems manager. We worked well together, and I learnt a lot about management from him over the following years. I was responsible for recruitment, among other things, and we gradually expanded the department to about one hundred and fifty staff. I became quite good at recruiting technical staff, but as I was to find out later on, I was terrible at picking good clerical and sales staff! When we had a major recruitment drive I would organise teams of senior staff to interview applicants in a two-level interview system.

Until the time arrived when all of the old systems had been replaced the old computers had to be kept running – and the old software. There was only one programmer left on the staff from the old regime. He realised his unique situation and threatened to leave. The company was absolutely reliant on him to maintain the old software, he was the only one with the ability and the knowledge. He blackmailed the company to give him a large sum of money to stay. The company paid up, and he stayed, but he was gone the moment the old computers left the building. This demonstrated beautifully that you should never get into a position of being reliant on any one person.

The next phase of my project was to install optical mark readers (OMR) at head office to read the part of the assembly slips being returned from the depots. This, if successful, would eliminate the costly, slow and error-prone key-punching of thousands of documents. I had seen these machines at an exhibition, saw that they would be ideal for our application and I had designed the assembly documents with this in mind. We had to train the assembly staff to mark the documents, but this proved to be no problem. The machines were delivered, installed, tested and implemented without any trouble. Virtually overnight we saved the keying of many thousands of documents each week. Every computer department then had a 'data preparation' section, usually staffed exclusively by women. Their job was to keypunch all of the data used in the computer applications, taken from manual documents. We had a section of around twenty or so young women, and slowly made inroads into the number by advances such as this one. Terminals on desks had yet to be invented.

It was around this time that I was invited by ICL, the computer manufacturers that made the 1900 series machines that we were using, to give a presentation to one of their customers. The board of directors of Massey Ferguson tractors were having a one-week course at one of ICL's country houses that were used for training and education. This one was down by the river Thames, Hedsor House, in a rather beautiful setting. Apparently the directors had asked for a couple of talks from customers who were using computers in a big way. I found myself in the company of a senior manager of BOAC (British Overseas Airways Corporation) – just the two of us were giving the presentations. The BOAC man was on first, so I listened to his talk which was polished. He used some classy visual aids and photographs as he told his audience how BOAC were using computers to schedule flights, check in passengers and presented a vast amount of data. I was a little nervous, following on as I was from this rather elated gentleman and a rather elated subject. I began my talk by saying that I was going to bring them down from the clouds to ground level, and told them that the very same year that the mansion we were in was built – 1869 – that a Mr J.J,. Sainsbury had opened a shop in Drury Lane in London to sell groceries. (I had happened to be reading the brochure that described the history of the house just that morning!) I went on to tell them a little of the Sainsbury history – how John Sainsbury opened other shops and had a large family of eleven boys. He wanted a shop for each boy to manage, but wore his wife out and soon ran out of boys! It all went down rather well with the audience and I quite enjoyed it. This was the start of a series of talks that I gave over the next few years, and was my introduction to the serious business of professional speaking. I found that I enjoyed it and I rather enjoyed the surroundings of elegant country mansions.

I was working hard and frequently was late home, travelling by train and tube from St Albans to Blackfriars each day. We decided to ease this problem by moving house, and moved to Oakwood, near the end of the Cockfosters tube line. This reduced my travel time to just over an hour on the tube each way. We bought a detached three-bedroom house that was close to a small lake and very close to my old college sports pavilion. Again I spent time on the house and over the next year or two I decorated it from top to bottom, installed a washbasin in Lucy's bedroom and had central heating installed. Lucy went to the local school but she became a problem child for a while. It took us a little time to work out that she was bored – she was almost reading before she went to school – having a librarian mother meant that she was read to and had lots of books around her – but the school was not able to move her on fast enough. After much discussion with the school and after some serious heart searching, we decided to send her to a private convent school. We were staunch atheists but the school, run by nuns, got very good results and had a high reputation. Lucy did well there and we endeavoured to join in the parent-teacher events. We always felt that we were never really accepted into the 'in' crowd as we didn't attend church and were one of the very few non-Catholic families there. The nuns though were very friendly and kind. Lucy, I am pleased to say, was not converted. She attended ballet classes on a Saturday morning and I shall never forget the smell of twenty sweaty little girls that greeted me when I went to collect her. She was able enough but she stopped going after a couple of years. We also bought a piano and had a teacher come to the house to give her piano lessons but Lucy bucked at this and played up until the teacher told us that we wasting our money. Later at secondary school we bought her a flute and she had more success with this instrument.

Our beloved basset hound died, and was replaced by another, Clarence. He was even more of a character than Calamity Jane, so much like the cartoon basset, Fred. We always made new friends when we took him for walks. "Hello, Fred" strangers would say and start talking to us. I took him to dog training classes and at first we (I say that advisedly, as in reality it's the owners who get the training) got on quite well. We graduated from class 1. Then he started to play up. He would 'stay', 'sit' and walk on a lead but getting him to retrieve was something else. He just sat and looked at me when I said "fetch", pointing to the wooden bone. "Encourage him," said the instructor. So off we went, up to the bone. "Pick it up!" I said. Clarence looked at me and did nothing. I was on my hands and knees almost picking up the bone in my mouth when I looked at Clarence. He was looking at me and I'm sure there was a smile on his face. He never did pick up the bone, and we were politely asked to leave the class when he became bored and started attacking the other dogs for a bit of light relief.

At work my major project was yet to come. The actual quantity of each product to be sent to the supermarkets was determined by a large, clumsy computer system that forecast what would be sold of each commodity in each shop. This was a complex business, involving a costly and laborious stock-take each week and a very involved computer programme – that was always wrong! The local variations from store-to-store, week-by-week were of course impossible to predict, no matter how clever the forecasting system was. Consequently stores would be overstocked with some items and out-of-stock of others. The directors searched for ideas. I and several of my colleagues went to France to look at the hypermarkets there and we visited several of the large companies to determine how they did things. None of them seemed to have any systems that were an improvement on ours. Then slowly the idea emerged that stores should no longer have a back-room stock. The stock should all be on the shelves for the customer to see. The space saved in the store room would be better-used as selling space – one common complaint was that Sainsbury was so successful that the stores were overcrowded with customers. (It was the biggest and most successful UK supermarket at the time, and boasted the highest turnover per square foot of any retailer in the world.) The other plank of the idea was that the branch staff should determine how much of any stock item they should have delivered, based on what they had left on the shelf at the end of each day's trading. Key information on the shelf would tell them what the space would hold. We would require a simple way of assessing the quantity of each item, placing an order for that item in the branch and somehow getting the order to head office for processing. The order would need to be processed in minutes, linked into the picking note system and sent to the distribution depots for subsequent assembly, before being delivered to the store, all within a few hours, so that the store had the goods the following day. It was what we now call a 'just in time' system. These were radical ideas, and ones that thus far were technically impossible to achieve. We had the means to produce the assembly data and transmit this information to the depots, but that was all.

A steering committee for the project (the direct stocking system) was set up, with an American consultant as a sometime member, and senior staff from the key departments involved. I was the sole I.T. representative and the most junior in status on the committee. The others were senior managers in charge of the retail and distribution departments. My work to date had given me the necessary knowledge of the retail end of the company and I had the confidence of the key managers. I was to advise the committee on what could be achieved and to lead the computer systems team to make it happen. A key issue soon emerged – the orders were to be placed at the branches by staff walking around the store and there was no hardware on the market remotely suitable for this task. All input devices were keyboard-based, and I ruled this out from the beginning, so we called for ideas. We received more than fifty ideas from companies and private inventors, including some weird and wonderful ones. Despite my saying that keyboard input was not a runner, ICL produced a glossy proposal centred on the idea of mobile keyboard devices connected by trailing cables to the electricity supply. The most promising idea came from Plessey, who had the concept of what became the barcode label that could be read by the swipe of a reading 'pen'. I liked this idea immediately and thought that it could work. This was a strongpoint of mine, I seemed to have a good notion of what would and wouldn't work in a situation. It seemed to be intuitive to me, but I wasn't always able to explain to others why an idea would or wouldn't work! I would have to sit down and work through the logic to argue it out, but I often wasn't willing to do this. Some of the crazier ideas had their proponents so I ended up arguing it out anyway with some of our senior people before we could knock them off the list of options. I quickly focussed on the barcode idea and spent much time with the development engineers, going to their site in Poole, Dorset a number of times. My engineering background proved to be most helpful here. Two development engineers arrived one day to demonstrate their product. It was a 'bread board' model, that is, it was constructed in the research laboratory and was just intended to demonstrate the principles. The demonstration took place in the computer director's office. The reading 'pen' as they called it, was a large thing about ten inches long that rested on the telephone cradle of a dial-up telephone handset. It did indeed read a barcode, that was itself about six inches wide, and the number represented by the code appeared on an electronic display panel on a large box. The director seemed impressed. "Of course," he said, "it would have to be much smaller." They agreed that it could be. "And", he said, "it would have to be tough", and with that he rubbed the pen very roughly back and forth across the barcode. There was a flash, and sparks and smoke emitted from the box of electronics, along with a strong smell of burning. The engineers looked crestfallen. "Oh well,' said the director, "these things rarely work properly the first time!" He gave the go-ahead for us to continue the investigation along these lines. Plessey came to us to ask for development funding but Sainsbury's turned this down saying that we were a supermarket not a development business. That was only partly true – we were into software development and the hardware development was essential – as we were on the leading edge. Plessey took the risk and funded the thing themselves, and were eventually rewarded when it all worked and we installed some five hundred of the devices in the stores and continued to buy newer models for many years.

Before that could happen we had to develop the system software. The ordering data, consisting of a product identifier and the quantity required (just one or two numerals), was to be recorded using a small key pad onto a cassette tape at the branch – about one thousand items from each branch each night. We minimised it further by having the system assume a quantity of one (pack) if no quantity was input. This applied to some half of the orders. This relatively low quantity of data was OK for public telephone lines. One issue was that the whole operation had to be foolproof as it was to be used by hundreds of people in the stores and it would be a changing population, thus little or no skill was to be needed. We worked hard at this, and came up with a system that either read the code correctly, or didn't read it at all – and told the operator so that he/she could key in the code. The barcode itself was designed to be self-checking and could be read either forwards or backwards. (For you numerate people, it contained an extra binary digit that had to make the total always an odd number). I wasn't happy with just this, I wanted belt and braces, so the reference number itself that was barcoded also contained a self-checking number. (Again, if you want to know, it was a modulus 11 Cyclic Redundancy Check). Reading the barcodes – about one and a half million reads a week – never produced any errors that we could detect and the barcodes very rarely had to be keyed in. From a systems viewpoint – very neat all round, though I say so myself. On top of this the usual self-checking and correcting system applied to the transmission of the data. When all the orders were recorded, the devices – for there were to be two in each store – would be plugged into a transmission unit, the data would be transmitted to Blackfriars upon a phone call from Blackfriars and we would have to collect it and process it then pass the data to the system I had introduced to produce the picking documents that would then be transmitted to the depots. This is where the speed of transmission was really important, as the depot transmission load was high. The data had been expanded to include product descriptions, physical address of the product in the depot and so on. Management summaries also had to be transmitted. We also had to set up a printing process to produce the barcode labels, and a team to allocate these and manage what would be an ongoing process to distribute these to the supermarkets to be set up on the shelves.

In the meantime we ran trials on the concepts. We installed a system into a couple of stores with a rough and ready mark-sense card system to collect the data. The data was called out over the telephone from the branches to operators at Blackfriars. It worked brilliantly, so much so that the directors insisted that the system be expanded immediately! The branches actually received the goods that they wanted, and the labour cost in moving stock from the warehouse to the front of the store was eliminated, saving much money. The trial system had many weaknesses – it was only intended to last a short time for one or two stores – but we had to expand it while we developed the full solution. At its height we had about fifteen part-time telephone operators at Blackfriars taking the evening phone calls from sixty stores.

The first data collection units were at last available. The unit was housed in a trolley with the tape unit, keyboard and barcode reading pen. It was powered by two lead-acid car batteries and was quite heavy! It cost about £5,000. Improved battery technology was just around the corner, but this was the best available at the time. We had developed and tested the necessary software and checked it all out. The depots had prepared the reorganisation necessary for the assembly, loading, despatch and delivery of the goods within a few hours – I had been a member of a depot reorganisation committee to do this. The first store to pilot the system was in Holloway, north London. With some trepidation I introduced the part-time employee who was to place the orders to his new tasks. His full-time job was a truck driver. He listened, took hold of the equipment, and trundled off. After half an hour he was back. "How was it?" I asked. "Fine," he said. "No problems." I was much relieved!

There was one change we introduced as a result of the trial. One evening a cleaner unplugged the transmission unit from the power point to plug in her cleaner. The transmission had to be ready to receive a call from head office that would then trigger the unit to send the data down the wire, all automatically. The unit was left unplugged, no data was received at head office and no goods were sent overnight.

With this system this was a disaster as the store would be woefully stocked the next day without an overnight delivery, and Sainsbury's placed great stress on having a minimum of out-of-stock items. We subsequently ensured that the transmission unit was permanently wired in. This is why we do trials. The system was highly successful and after a little fine-tuning we installed it into all the new supermarkets as they opened up, and 'backfilled' the others that were using the temporary system. It took a couple of years to get all the two hundred and fifty supermarkets onto the new system, the branch hardware alone costing almost three million pounds. The managing director called it the most significant change ever made in the one hundred + years history of the company. The labour savings alone in the branches exceeded all projections, but the biggest benefit was the improved stock levels in the branches, as minimum stocks were held consistent with not actually running out of stock – the ultimate retailer's dream. And the supermarket manager had control over the branch stock for the very first time. Over the next few years the units improved as the engineers had access to the new battery technology arising from the space research programmes. The first improved unit was in a shoulder bag, then it became a hand-held device. Now of course barcodes are used world-wide, but this application was the world-first commercial application. I received a number of enquiries from companies who wanted to know more about the technology and one job offer from a business that wanted my expertise immediately! I turned this down as I had no interest in the company concerned. (Just to be quite clear, the barcodes that are scanned at the checkouts these days came along later, after the UPC (Universal Product Code) was agreed and came into wide use. Similarly the barcodes on books, which is the ISBN (International Standard Book Number) came along later.

During the implementation I was asked to attend a directors' main board meeting. They wanted an update on the project so I went along to the directors' floor with the founder's bust and along the rather swish spacious hall to the boardroom with my boss, the computer department manager. Mr J.D., Mr Timothy and Mr Simon as the Sainsbury brothers were known to the employees, were there along with the rest of the board. I gave a brief report as to how it was all working and on the installations in the branches and was quizzed on a few points but there was no drama. My boss didn't say a word – he couldn't really as he knew little about the project! The directors were all very pleased with the results, and happy that no disasters had occurred. All too often with large IT projects new systems are abandoned or cause a major disaster, and this remains so to this day. My team had put in much effort to ensure that not only did the software and hardware work, but that the system was understood and could be easily used by the branch and depot staff. As a token of recognition the IT director invited me and my immediate manager, John, to lunch with him in a director's private dining room. A butler attended us and it was an extremely good lunch accompanied by suitable wines and cigars (I used to smoke them then). I didn't do much work that afternoon.

The new system's success caused me to experience the directors' impatience when I was asked to ramp up the implementation. We were adding branches, product groups, and new features and my plan - to take it gently, came under attack. The savings to be made by the system were huge, and the directors were like kids in the candy shop. They wanted it all - now. "Why are you so cautious?" my director asked. "Why not implement the whole system immediately, all branches that are ready to go, all depots? That's what the board is demanding."

I explained patiently that things might go wrong. It was all new - the technology, the software, the manual operations in the supermarkets and the depots, all tested, but not altogether in a 'live' situation. By implementing in steps, I explained, if anything went wrong then firstly the damage would be contained, and secondly, we'd have the whole team on alert, watching everything, ready to pick up any issues that arose - and we would have a good idea where to look for the cause of the problem.

"But what will go wrong?" he asked. "Don't you have confidence in your work?"

"If I knew that" I responded "then we'd be fixing it. I just don't know. What I do know from experience is that in situations like this something always goes wrong. It might be a tiny software issue that has slipped through, it might be the new hardware, it is often a human issue that we haven't foreseen, a misunderstanding, a wrong instruction, a process that just doesn't work, but there'll be something. There always is. And it could lead to disaster. With my plan, we don't move to the next phase until we're happy with everything".

I went on to describe the chaos that might result if we went in with one total implementation. We were after all, talking of about 1.5 million product orders per week from the 250 supermarkets with five distribution depots and thousands of assemblers, dispatchers, truck drivers and branch staff involved. The possibilities were endless - wrong orders being produced, no picking instructions getting to the depots, no goods or the wrong goods being assembled and distributed, and the final result - the Sainsbury crime of all crimes - supermarkets out of stock. He visibly paled. I had convinced him of the rightness of my plan.

There are a number of possible implementation strategies that one can adopt and all experienced business and system analysts are taught them (as I was to in a later career move). Senior managers, however, as I found out here, are an impatient lot, and often force bad implementation decisions on their I.T. providers. I was not prepared to do what my director wanted, under any circumstances, as it was totally daft. My next step if he continued to press me, was going to be a written report, laying out the risks, which I would ask him to sign. He wouldn't have, I'm sure, but my final step might have been to resign rather than lead the charge to disaster. I'm glad it didn't come to that.

Businesses though, have no corporate memory. People move on, new managers appear, decisions are made without the past experience to guide them. I had sat on a committee to examine the question of automated depots in the 1970's but we decided against it on the grounds that when a breakdown occurred (hardware or software) distribution would cease totally as there could be no fall-back position. And as we were installing a "just-in-time" system this was unacceptable.

In 2004 that is exactly what happened, only there was not one but four depots suffering from frequent breakdowns! New technology was introduced into four new automated distribution depots, (each costing one hundred million pounds) and using bar-codes, all at once. It all went wrong and distribution ground to a halt. What a classical and expensive error to make! Whether this was the fault of the Sainsbury management forcing the pace or the I.T. provider not having the experience to know what to do, I don't know. I have no inside knowledge. Ultimately, the decision for these things is in the hands of the client. The supermarkets kept running out of stock and Sainsbury had to abandon the four automated depots. It was a disaster on a massive scale, all so unnecessary. At the time of writing they have yet to recover their number one spot in the UK supermarket hierarchy.

My system attracted some attention in the technical world and the retail world and I gave a number of presentations on it to the British Computer Society, working up a standard talk, using a demonstration unit that Plessey had built for me. At one talk at a college I was just getting into my stride when the door burst open and an irate man charged in. "You've got my projector," he tried to say, but he was in such a rush that he ran into the whiteboard that I'd set up and ended up on the floor, tangled up with the whiteboard. There were some muffled laughs. "And for my next trick," I said... After giving a talk at a commercially-run conference (when I received a couple of very nice bottles of wine for my troubles) I was asked if I would do one for the International Federation of Retailers. I must have been preoccupied when the phone call came, for I agreed and immediately forgot about it. Some weeks later the chap rang me to ask me about progress. I quickly tried to gather my thoughts. "Yes", I said "all's well." "What do you want for visual aids?" he asked. An overhead projector would be all I wanted, I replied. "Oh no!" he said, "that won't do!" I asked why not. "The audience won't be able to see if you use an overhead projector". I was getting a nasty feeling about this and then asked him, "How big is the intended audience?" He told me "At least four hundred" and I went silent. This was going to be different, I had become complacent and had forgotten to ask the basic questions. The talk was to be held in the Mayfair Theatre in London, and was clearly going to be much bigger than anything I had done before. I had some professional 35 mm slides made up to illustrate the system. In the event it went well, with many of the audience wearing headsets to listen to my talk in French, German and other languages.

At another talk in Amsterdam I booked into my hotel and asked where the gym and pool were. They didn't have these facilities, they explained, but guests were welcome to use the commercial gym just down the road. I went out into the icy evening down to the gym. It was strangely deserted but I found the changing room and got changed. It was then I noticed a pair of ladies high-heeled boots by the locker. I rushed to the door looking for the sign that would have told me I was in the wrong changing room. No sign and no other changing room! I gingerly went out and found the pool. On walking in I found that the mixed participants were all nude! There was a bar, with people sitting around on bar stools, quite naked, chatting and drinking. I slipped back to the changing room and thought quickly. This was a new experience for me and well, I was British after all. Although I had lived through the swinging Sixties, they had largely passed me by somehow, in fact I felt closer to that generation that was covered in goose fat and sewn into their underclothes for six months of the year. After a few moments' thought I decided 'when in Rome... ' and went back to the pool suitably unclothed. I felt like John Wayne entering the Wild West saloon bar as I pushed open the door but of course no-one took any notice of me. I'm unsure whether I was relieved or insulted. After my swim (when I was very careful not to brush up against another body) I was taking a shower in the deserted open shower room that had no cubicles, when two nubile, nude, shapely girls came in, stood either side of me and showered, chatting nonchalantly in Dutch as they did so. I wasn't sure where to look and turned on the cold tap. I never did establish how people paid for their drinks when they had no-where to keep their money.

So it was that when I was asked to talk in Paris to some five hundred international retailers I was well-prepared. I was following two IBM speakers and this had me a little worried, in case I was going to look a poor comparison. However, they gave a rather technical and boring talk, and many of the audience were dropping asleep by the end. So by the time I went on I could have talked about washing powder and still have been welcome! I began the talk in my bad French, having found a French woman in Sainsbury's who translated my introduction for me (which I had written out in full) and she had coached me with my accent. I then continued in English, no doubt throwing the interpreters. Knowing my subject well I had only to make the occasional glance at my notes and I now had excellent photographic slides to show. I had to pause while the interpreters spoke in their various languages and the audience listened on their headphones. The talk went down well, and at the panel session that followed, I got all of the audience questions, so much so that I was a bit embarrassed and tried to draw in the other speakers. The chairman of the conference was a director of Renault and he had arranged a day where the speakers could get together for "le preparation". I had flown from Heathrow to Paris and back the same day for this, but the whole day was merely an excuse, I thought, for an extremely good meal in a very expensive restaurant. We had hardly mentioned the conference at all! After the actual talk my party (I was accompanied by the computer manager and our ICL representative) went to dinner at the Moulin Rouge night club with their nude floor show. It was free to go in, one just paid an exorbitant price for the champagne! I didn't pay of course. ICL paid for the whole trip as Sainsbury's used ICL computers and they got a good mention in the talk.

Although my new system catered for the majority of the goods that a supermarket sold – things that came in boxes, packets and the like – there remained a range of products that were not so easily handled. Meat for example, fish, deli items and fresh produce. My new system was working so well and saving so much money that the directors wanted it to be extended to all product groups as quickly as possible. So we tackled these in groups, devising special system options to enable the overall branch system to accommodate the special criteria of these products. When I began work on one group I interviewed the senior manager involved. He had about fifty staff and it was quickly apparent that if I was successful and we managed to automate the work then most of the staff would become redundant. I asked him how I was to proceed as I would need to talk to the staff to establish what they did and how they worked, and they would want to know what the future held for them. He told me to play down these fears and to say that all would be well. I answered that this was not a good policy and that the staff weren't fools, they would know what was going on and I would not have their confidence and co-operation. I explained that much involvement and co-operation would be required for a successful system design, testing and changeover from one system to another. When he refused to accept this I had to say that I was not willing to proceed along these lines, and that he had to come up with a policy and declare it to the staff before I could begin work. He was reluctant to accept this but I insisted, a bit odd really as I did not have senior managerial status but I did have a status by reputation by now. And in fact if he had not played ball then I would not have hesitated to take it up the line to the top if necessary, as I was confident that I would be backed up all the way. He did come up with a policy whereby alternative positions would be offered or a redundancy offer would be made to the staff, no doubt after consultations with his hierarchy. The staff accepted the agreement and I was then able to proceed with the detailed work. In the event forty or so of them did either move on or take redundancy. I was just amazed that a manager could be so naive as to think that he could keep his staff in the dark during such a process as we had to go through with them. This was still in the Seventies but the inexorable march of automation was already well underway.

The system was extended to non-food items that were now appearing in all the supermarkets. I was in a branch in south London checking out something when I was approached by a young woman assistant who wanted to discuss an issue. "It's about the knickers," she said, "let me show you." "What, in the middle of the supermarket?" I stuttered. She dragged me off to the ladies' clothing section to show me the undies range. "These are not the style required now," she said spreading out a sample or two. She took no notice of my protestations that I was a computer nerd in the I.T. department with no authority over the styles of ladies' knickers on sale, assuming that as I was from head office I could do something. She continued at full gallop. "These are for old ladies," she told me, "and most of our customers want something different." Now even with my limited knowledge of these things, I had to admit that the waistline looked to be rather close to the armpit and the leg line too close to the knee for most females (or males for that matter). There was sufficient material there to make a pair of curtains. She insisted on telling me what was required, which was something closer to two small triangles and a piece of string. I promised that I would pass on this information to the appropriate authority at head office. Back at Blackfriars I found the buyer of ladies' underwear – a man, to my surprise, and explained the problem to him. He seemed to take notice and thanked me. I sent a note to the young assistant to tell her and pass on the thanks. "I trust that your knickers will now pass muster," I wrote. Whether the style of undies was brought up to date I can't say, but I know that Marks and Spencer came badly unstuck a year or two back when they failed to keep abreast of the latest fashions and lost swathes of customers. It's a tricky business, fashion, but even I could have made a better job of buying undies that that fellow. I thought that maybe he lived with his granny and didn't get out much. Incidentally, do you know why we say "pair of knickers" when there is clearly only one garment? In case you don't, originally a knicker was one garment which was drawn up one leg only, (hence "drawers") and fastened around the waist with a cord. Another made up the pair. No crotch, presumably making it easier to have a pee when one might be wearing petticoats, hoops, bustles and tight corsets, making it nigh impossible to bend, yet alone remove anything. Now you know.

After the dust settled on this project things became a bit more humdrum. The project had been leading edge and most exciting to work on. The use of barcodes was a world first, and our data transmission systems were also advanced for their time. The system had attracted lots of attention internationally and of course I had been highly motivated. I now had two project teams to manage and lots of software maintenance and changes to carry out. I was also responsible for recruitment, and I interviewed many scores of applicants over the years as the department grew to over one hundred people.

One day the computer manager came to see me in a bit of a state. It was getting near to Christmas, a mad time for the business of course. He had come straight from a directors' meeting. They wanted a systems change made that would affect the distribution operations that very same evening. He told me what he wanted to be done. "Do this yourself," he said "immediately! Let me know when you've done it." He was clearly feeling and looking as if he was under some pressure. I sat at my desk and considered the task. It was something that could be done with a procedural change in the computer operations, and with a tiny programme fix. This sort of quick change was always something to be avoided, as it was commonly experienced that these "tiny changes" often go wrong in the heat of the moment, with no time to either think them through or to test them properly. With the huge volumes of goods that were involved – just one supermarket might receive ten huge trucks full of goods every day and as there were several thousand men involved in the assembly operations in a number of depots in a time-critical environment, we couldn't afford any mistakes. It quickly became clear to me that what I'd been asked to do would not work at all. In fact, it would generate an enormous pile of reports that the branch staff would not be able to handle or action, just the thing to avoid at Christmas! The computer printers would be working for hours just to print everything. I went to see the manager and tried to tell him that it was not a good idea. He exploded. "The directors have given a direct order," he said. "Just for once do as you're told without raising issues!" He refused to listen to my protestations. After more consideration I went back to see him again. "Pete," I said, "this is crazy. It won't work at all..." "You're fired," he shouted. "Get out!" I wasn't worried by this, I put this outburst down to stress and ignored it. But I was in a dilemma. My direct manager, John, was out, so I had no-one to consult. I rang my director and explained that I urgently needed to talk to him. He had been at the directors' meeting and so understood what I'd been asked to do. I went up to the directors' floor and to his office and explained the disastrous situation that would result if I went ahead as requested. "What do you recommend?" he asked. I checked my understanding of the directors' intentions. "Then this will achieve it," I said, outlining a plan that I had constructed. He pressed a buzzer and called in his secretary. "Ask ... to pop in please" he said. Two directors appeared within a minute. "Tell them what you've just told me, Derrick," he said. I did. They listened and asked a couple of questions. "Fine," they said, "please get on with it". I did, and it worked. I didn't know how to deal with the computer manager after this. In the end I just knocked and popped my head around his office door. "Pete," I said, "what we did worked as planned." I left it at that. He knew what had happened but neither of us ever mentioned it again. This episode was a great example of senior managers (directors in this case) getting into areas that they don't understand. When there's a requirement or a problem, one should always consult the experts – not by asking if a certain solution will work, but by giving them the problem – and the objectives –

and letting them come up with solutions – that's what you are paying them for! I always tried to do this myself when later I was in a similar position. Your staff will always know more about the details than you do and will probably be very inventive.

During this time Jennifer and I had moved yet again. This time we it was a mile or two up the road to Cockfosters. We bought a detached house from Dave Davies, who, with his brother Ray were the frontmen of The Kinks pop group. The place was a shambles. The rear garden was chest high with grass and weeds so we had no idea what was there. The hall was bright red – carpet, walls, ceiling. The walls in fact were covered in hessian, soaked in red paint so that when I pulled it off it came off like sheets of wood – taking off the plaster as well. The sitting room was orange and brown, and the utility room was orange and purple. The only room half-decent was the kitchen, where I imagined his wife/partner must have exerted some influence. There were holes in the ceiling of the sitting room where hooks had suspended cables, as this had been a group practice room. A dartboard had been fixed to one wall, and the wall was full of holes from the darts. Some rough timbers covered the wall nearest to the neighbours, and this had been intended as soundproofing. From what I heard from the neighbours it hadn't been very effective. There were broken lead-lights, poor condition wiring, and the whole place was draughty. There was a third floor with a playroom and a small room that I intended to use as an office. The main bedroom had an en-suite so although it was a mess, it had potential, as the estate agent had said. We moved in and I began to work on it. It took two years or more as I had to largely rewire the place and do some substantial work on it. But it came good and we were very comfortable there.

It was during the move to this house that I put my back out as I was dismantling my much-valued workbench. I picked up one of the heavy sleepers that formed the benchtop and turned... it felt like my back had broken in two and I was in agony. This injury became a permanent fixture and I have had to live with it ever since. I have learned to manage it and to minimise the risk of a recurrence but I still have an episode once a year or so that lays me out. It takes a few days to recover with some appropriate exercise and massage.

The family now included a cat as well as Clarence. It was a good ratter which was just as well as north London had plenty of these horrible creatures. The cat would sometimes catch a large rat, drag it into the house through the cat flap, then play with it. We were just leaving the house one evening when a movement caught my eye on the curtain rail. A large rat sat there, looking at the cat! I had to knock it down and kill it with a broom handle. On another occasion the cat let a rat go in the kitchen, whereupon it wisely hid behind the fridge. Tucking my trousers into my socks (I had this notion that rats would run up your trouser leg) I shut myself in the kitchen with Clarence and the cat, rather hoping that between them they would do the required job. I dislodged the rat from its hiding place when it ran around the kitchen looking for an escape route. Clarence just sat there watching but ignoring my entreaties to do something. The cat similarly sat and enjoyed the spectacle. I was forced to chase the rat myself and wallop it with the broom handle whereupon it came to a halt when I could get a good swipe and despatched it. I've always had a certain fear of rats ever since.

The Sainsbury experiences were teaching me something about management, especially managing and motivating technical people. I did a couple of courses and became interested in psychology. We all have egos and want to achieve, but getting technical people to work well together I found wasn't always easy. Many technical people are nerdy and some find it difficult to communicate in writing or speech. In fact, the more I found out about systems development work the more I realised that project management was a key factor, and that people management was an important part of this. My theoretical knowledge was based on what I had studied from my management diploma, and now more began to fall into place. Some of my own weaknesses became more apparent to me and it became more and more obvious that the technical know-how was sometimes less important that the people skills. One encounter gave me the opportunity to try out some people skills. One of my systems analysts, a very bright young woman, had a reputation as an abrasive character. One of the computer operations shift managers came to see me one day to complain of her behaviour. "She rubs everyone up the wrong way," he said. "None of my staff want to speak to her. She's real trouble." I promised to see what I could do. I met her in a private room. "How are you getting on with the operations people?" I asked her quietly, after we had discussed her project. She hesitated. "Not very well," she said at last, "they're a bunch of wankers. I can't get them to co-operate." "Is this the view of your colleagues?" I asked. "Do they all have similar problems?" She thought for a moment. "No," she said, "I don't think that they do." "So," I said, "you're telling me that you are having problems with a bunch of people but no-one else is?" There was a silence.

"Does this say anything to you?" I asked. Silence again. "I suppose," she said at last, "that the cause of the problem might be me." I said nothing for a time while she struggled with this. "Do you think that you can do something about it?" I asked. "Yes," came the reply after a long pause, "I think that I can.". A week or so later the computer operations chap caught up with me. "What's happened with _____?" he said. "She's all sweetness and light and we're getting on fine with her now." "I don't know what's happened," I said, "but I'm pleased it's working out." And I was. It seemed that people are often willing to be honest about their weaknesses and, when faced with them, they will do something about them – but only if they want to.

At some point I was asked to join a high-powered committee that the directors were setting up to improve efficiency. Apparently Marks and Spencer had done something similar with great results. It was dubbed the "paper-chase committee" and was given the task of reducing the administrative work in the branches. Many paper systems were in use and more were introduced each time that the head office people wanted information. Some of the systems were rather clumsy and amateurish, with multiple copies of forms being sent to different departments. But Sainsbury was different to M & S – Sainsbury had been using computers since the early days and had made great strides in the systems that handled the ordering and distribution of goods, both into and out of the distribution depots, while M & S had been slow off the mark. Nevertheless there were gains to be made in Sainsbury, but I felt that the committee was not the best vehicle to achieve the required results. I was the only I.T. person on the committee and the only non-senior manager. It was a repetition of the branch ordering steering committee that I'd sat on. We made some improvements and reduced the number and complexity of the forms in use. More savings would have been available from new computer systems I felt, but they would have involved new projects being set up and in particular would require computing power and good data communications with the branches. At the time the personal and small computers had yet to see daylight and my highly-specialised ordering system was the only way of getting data from branches, so these limitations were major stumbling blocks.

I attended many meetings at Sainsbury's, some of them having twenty or more attendees. Meetings can be terribly boring, time-wasting affairs and this is common still today. One senior manager was a brilliant chairman who kept his meetings short and to the point. If you were actioned to do something, woe betide you if, when asked for a report at the next meeting, you had not fulfilled your task! He started his meetings on time, having a clock in front of him. If someone came in late, he would make a great show of thanking them for deeming to turn up, by pointing out that the meeting had been running for x minutes, and how grateful we all were for their esteemed presence! You were never late a second time! At one meeting a main board director had been asked to attend. When he didn't show, the chairman asked someone to make a call to him. The director could be heard saying that he wasn't able to attend. The chairman took the phone and told the director that he'd agreed to come, that he was critical to the decision that had to be made and that he had to get himself there right now, or they would both be fronting up to the M.D.! The director quickly appeared. This manager clearly was a person with clout, and he used it – along with some colourful language – when he had to. He was actually quite a charming man in many ways. He had it right about meetings. When the cost of the attendees is taken into account, they can be very expensive time-wasters. Meetings can also be fun, and this chairman took every opportunity to make them so. I learnt a lot from him, and tried to apply these lessons when my turn came to chair meetings.

At this time most software development performed internally (as opposed to contracting an outside company) was done without a charge to the client. That is, the computer staff and all the costs of running the computers were an overhead on the business. Many companies operate this way to this day. As it became clear that computer systems not only saved money and time but made the impossible possible, then management wanted more and more development work done. But they had limited resources. A fight would often develop between directors and senior managers as to which of their pet projects was more important to the company. Now a cost/benefits analysis is the way to settle this argument but that also takes time and costs money – and it is only as accurate as the assumptions and estimates made i.e. not very! I learned quickly that when a resource is free then the demand for it is almost infinite! Managers would make demands and sometimes the one that shouted loudest got the resources – because there was no pain for them, no cost on their budget. I experienced a good example of this when a senior Sainsbury manager made unreasonable demands, as I saw it, when we were engaged in automating their allocation process. His department allocated fresh produce that was often in short supply. The demand from the branches might be say, for ten tons of strawberries but only seven tons might be available. There might be many reasons why a simple share out on a proportional basis would not be the way to go, and various situations were discussed. We established the simple formulae that his staff used. But he wanted more formulae to deal with these unusual situations. After much work and imagination we came up with a further nine formulae that they thought that they would need. Some were quite improbable but he insisted that they were all wanted and I had little to argue against it although my suspicions were strong that he was over the top. The design, coding, testing, and implementation work was considerable, but we did it. Unknown to anyone I inserted a count in the code that checked how many times each formulae was used. After twelve months I checked the counts – only the one basic formulae had been used! The manager was just a little embarrassed when he was given this information. The only solution for this is a different culture – where the client pays for development out of a department budget, then they think very hard before asking for doubtful functions.

Without this 'charging out' culture another related problem that arises is the setting of priorities. Management may set priorities for projects, then at the next meeting, may change them. Some managers seem to assume that one can treat software development teams as if they are bricklayers. To change a project, you merely get the brickies to move to the next brick wall and start laying bricks. With a development team there is a considerable 'pick-up' and 'put-down' cost as staff have to become familiar with the issues, design, and each other, and the manager has to plan and allocate the work. When a project is stopped, restarted, or has the team changed there will be more time and costs absorbed with nothing to show for it. This is why when a project is late and over budget and management then throws resources at it in a desperate attempt to get progress back on track they will almost always be disappointed.

After a time I was asked to take over a project that was in trouble and this was the first time that I became involved with personnel and payroll systems. It was well behind schedule, was in a muddle, and the project manager had resigned. When I found out who was in the team I immediately requested that we have a project area as the people were scattered about various buildings and floors. I was told that this was impossible, there was no suitable space available. I argued strongly for an area where we could sit together, this being most important to foster any sort of team spirit. This was one of the few occasions when John and I had a disagreement. I had to say that I was not willing to take on this new role unless I got my way – and I got it. On the first day in my new chair I introduced myself to the senior personnel manager who, it transpired, had not been kept up-to-date with the project development. I asked what the issues were that this new system was going to fix, and why they were important. He explained these things to me and emphasised that the implementation date was an absolute – the whole structure of the retail staff organisation (who made up the majority of the sixty thousand staff – was being changed and there was a strict timetable of events to adhere to. He was most concerned to say the least when I could give no promises on anything at that stage. Amazingly, after a couple of days of meetings with the various personnel people who were to manage the system I found that I knew more about the environment and the issues than the rest of my team, who'd been working on the project for months! It was a case of the 'ivory tower' and poor communications with the client. It was quite easy to fix that bit but a little more difficult to make the required change in strategy. The main issue was a large old computer program that had to be changed. It had little documentation – quite usual even these days – and there was only one programmer who was familiar with this programme. When I looked at the specification of required changes I recoiled. "This is a huge amount of work," I said, "what's your estimate of the work required?" The programmer, an experienced chap and quite senior, said that he would have it finished within a few weeks. He'd already been working on it for a couple of months and I had no faith in his statement. Like most I.T. people, he was an optimist and he had a reputation for never meeting his estimates. I looked at the options and realised that currently all the eggs were in one basket, and we had no time for further delays. I explored another option – that of scrapping the changes and writing totally new programs to achieve the desired result, on the basis that starting from scratch was often simpler than trying to change the current product. Operationally, it would not be the most efficient solution but computer time was not the issue here. The software would only be run once a week, and it would not be time-critical. My programmer was most upset by this suggestion, seeing it as a slight against his competence, which of course it was, and he fought back. However, I persisted and found that we had a number of trainee programmers coming on board within a few days and we wanted work for them to do as a training exercise. I insisted that we respecify the changes and design a number of new programs. The dozen trainees would write these under strict supervision, following a strict methodology. This is what we did and the trainees were carefully supervised at every stage by a very good senior programmer that I co-opted onto the team. Most of the programs worked correctly on the first test run – something that was hitherto unknown! Meanwhile my senior programmer suddenly was absent one Monday morning. He had broken a leg while playing hockey and he was off work for some weeks. This didn't affect my schedule at all. The project went in on the scheduled time and the client was delighted with the result. I had learned that on taking on an existing project – especially one that is in trouble – that one must look carefully at the strategy being employed and if it's found wanting, to change it, regardless of the opposition from those with something to lose by the change. Sometimes this will make you enemies if they cannot be convinced of the rightness of your proposal and I made one or two along the way at Sainsbury's. I think that by the time I took my next job I had improved my powers of persuasion. Having more authority also helped.

One lesson I learned here was that I had a habit of assuming that everyone knew what I knew. Now this may seem strange and totally unrealistic to you and I don't know how or why this came about but it landed me in some trouble now and again. I would forget to explain my reasoning or actions, and be irritated when I had to explain them. I knew about selling ideas to management and users, and I became quite proficient at it, but with my colleagues I was deficient. I came to realise this when I hadn't explained fully why the barcode reading system was so specialised. Others had seen the potential for the data-links that we had installed to the branches but the system was only suited for my application. The reasoning was, of course, that the application was very large, it affected the very core of the supermarket operation and had special requirements. To have made it suited for more general applications would have compromised the accuracy and efficiency to an unacceptable extent. This was obvious to me but not so to others! There is a parallel in nature – a highly specialised bird or animal does very well in their usual environment but cannot cope well when that environment changes. I came up against a related issue when the computer director was asking questions about the development of the system. He liked to quote the "KISS" principle, (Keep it simple, stupid! – a rather not politically-correct acronym). At one stage he forbade me to include a function on the basis that the system was becoming too complex. I asked him what he actually meant – as to have a simple system meant more work for the users, but to have a system that was simple to use was to have a system that was more complex 'under the bonnet'. An automatic car demonstrates this – it is simpler for the driver but has more complexity in the workings. You can't have both! This time he was concerned about timing and that we wouldn't meet the deadline for the system. I knew better – although the additional function I was contemplating was not required immediately, it would soon be and then would be a huge amount of work to add it in. So I ignored him and added in the necessary software (at little extra cost) leaving it switched off until later. When it was required a few months down the track a simple code change to unset the switch made it all possible. Of course, when he asked me how long the change would take I pursed my lips and said that it entailed amending and testing many programmes – many weeks. When after one week I told him it was ready to go he said "There – I said that it wouldn't take you long!"

CHAPTER 6
Plessey

Towards the end of 1976 I was ready for a change. Sainsbury's had given me some really great opportunities and I had enjoyed the excitement of my projects and the new technology that I had introduced. (I notice that in 2013 barcodes were still used on Sainsbury's shelves to identify products, so presumably at least some of the essence of my system remains!) But now I felt that I was growing roots under my desk; everything was too easy, too familiar. And I was a bit frustrated at the lack of promotion – the next step for me was to a senior manager but there was only one available in our department, and that was John, my boss. I thought that I was ready for more responsibility, more challenge. I put out some feelers with a recruitment consultant and got several interviews, including one with the Plessey company, at Ilford, Essex. It was a large company with some sixty thousand staff, very similar to the size of Sainsbury's. It was totally different though with many manufacturing sites around the country (including the one at Poole that I was familiar with), and research and development establishments. I was happy that my engineering background would also help me here. The drawback, of course, was that people might misconstrue the situation - this was the company that I had been dealing with after all. In fact no-one at Plessey that I came into contact with had any knowledge of the link between Sainsbury and the small unit at Poole which manufactured the data capture units that Sainsbury used. There were a few eyebrows raised at Sainsbury though, and I had to wear that. I had several interviews with different people. The job on offer was that of information services manager (their title for I.T. manager/computer manager) for the Ilford site, which happened to be the head office. At the third interview I became a little impatient. "I've been here three times," I said. "I think that it's time for a decision. I do have other jobs lined up (I did) and I need to know where I am with you." They had an issue with me – I was unfamiliar with the operating system that they were introducing. "Do you want me to do the work, or to manage it?" I asked. "If it's the former, then I'm not your man, if it's the latter then I think I am." I got the job, with a company car, a large office, a secretary – and about twenty systems analysts and programmers to manage. This was the only time that I nearly came to work for any form of government. I was on a short list for the role of manager of the Kingston and Sutton joint computer organisation. I was told that I was the preferred candidate in fact. But when the time for the decision came, it was deferred due to some disagreement between the two councils. At this point I backed out of the running, as I had reservations about working for local government, having had close encounters with some, and this was typical of the issues that frequently arose – getting decisions! This was a good decision for me with hindsight as I'm sure that my style of working would not have suited the sort of bureaucratic organisation of local government.

So I accepted the offer from Plessey. The department head count was fifty so we were well down. I had been told that the department was in a bad way – and it was. Everything I looked at needed to be fixed. The morale was low and more staff were on their way out of the door. The department was a run on a profit-basis, that is, I had to charge my clients for all the work done, pay all the bills including salaries, overheads, rent, lighting etc and make a certain profit percentage on the capital employed. It was currently making a large loss. My clients – the Plessey businesses based at Ilford, plus other sites around the country – in turn were free to take their business elsewhere and buy computers, software development, everything from the outside market if they wished. This was an unusual concept at that time and was considered to be a tough business to run. It really was just like running my own business, and I had always had a desire to do this. Of course, I was required to follow the Plessey rules and regulations, of which there were many. Nearly every day the post would contain additions and amendments to the rule book. The secretary would add them to the huge binder which I dropped into the bottom drawer of my desk. I didn't open the binder for twenty four months. I had so many bush fires to put out that I had no time for the red tape. Every now and again my manager, Joe, a rather frightening-looking large bloke who looked as if he was the godfather of the mafia, growled at me "You know Derrick, it really is time that you stopped using the excuse that you're the new boy on the block and started following the rules. I've had another complaint that you didn't..." I nodded, but continued to ignore these entreaties for some time. I got away with it because I was beginning to show success. This wasn't particularly difficult because the previous manager had been such a dill. In fact they had had four managers in three years, each one doing a good job of stuffing up. No wonder that the staff had been leaving! If I could just do a few things right I was going to be bathed in glory! At least these were my thoughts when I took on the job, and I could now see why there was much hesitation at the interviews. Unknown to me the staff, cynics that they were (and who could blame them?) were running a book on how long I would last. I outlasted the most optimistic (or was it the most pessimistic!)

I was one of twenty information services managers, each responsible for the I.T. services on one of the factory sites dotted around the country. My site, being the head office site was the 'home' of five of the Plessey businesses, so the CEO and five managing directors had their offices there. It was huge with thousands of engineers and factory workers employed. The I.T. organisation headquarters was also on this site, so my department was always at risk of being looked over. My immediate manager was based at Northampton, so he would only visit me once every two or three weeks, he told me. He spent the first day with me but I really didn't see much of him after that, and he was little help to me with any issues. The next manager up the hierarchy, Joe, was at Ilford and I tended to have more direct contact with him. Above him was the I.T director, responsible for worldwide I.T., for Plessey had factories in the U.S.A. and Europe. My first few weeks were spent grappling with what the department was doing and what was needed by the customers. Each I.T. site was measured by various criteria and these were published monthly as league tables. For example, each computer operation failure due to software was recorded. My site, Ilford, was at the bottom of all of the fifteen tables! I had meetings with all of the directors and managers of my client businesses and established the beginnings of relationships with them. As the 'new boy' at first I was only listening and asking questions, taking note of their complaints of the past service that they had received. I soon realised why my predecessor had been sacked. One or two of the client managers were quite difficult to get along with, and were quite aggressive in their demands. All I could do at this stage was to map out how we would proceed with regular meetings to establish plans and to monitor progress. I also had to grasp what systems were in use and under development and how my staff were currently organised. Asking computer staff about their systems was asking for trouble. Few of them were able to explain briefly what the systems did or the underlying principles. They would all quickly launch into great detail, leaving me stone-cold. They had no overall diagrams to show, which I desperately required. I was in the position of so many of our customers who had no understanding of their systems. This of course was, and remains, a major communications stumbling block in I.T. – there is usually no easy way to bridge the gap between the technician and the user community. I had to avoid getting into the details – that wasn't my job and there was no time for it – so I quickly learned what questions to ask and would shut up my staff when they got into detail. Over time I managed to train them to be able to communicate the essentials.

At the end of my first week one of the senior staff popped his head around the door. "Fred is having his leaving party down at the pub," he said, "would you like to join us?" I accepted, and down to the pub we all went. Friday lunchtime pub sessions were rather a tradition with many computer departments and I had seen the results at Sainsbury's. People would roll back at 3 p.m. and very little work was achieved that afternoon. I had once been severely embarrassed one Friday when I had been asked, without notice, to give a progress report to my director, after such a session. I didn't want that culture here and so, at 1.55 p.m. I pointedly looked at my watch and said that it was time to be returning to the office. I got up and left the pub without waiting for them, and returned to an empty office. I watched the door, wondering if they were going to accept my lead. I had the advantage that they didn't yet know me, and sure enough they all trooped back within a few minutes. This set the tone and Friday lunchtime drinking was not a problem after this. Well, only once. A chap came late back and was not A1. I called him into the office and told him that I had to send him home and that he would lose the afternoon's pay. That was the end of this issue.

My company car was on order and until it arrived I had the use of a pool car. This was an old large Ford with very sloppy steering. Driving out of Ilford one foggy wet winter night during my first week I managed to hit a car at a junction. I took the damaged car to the transport department the next day where they supplied an equally old saggy replacement with hardly a raised eyebrow. Only a few weeks later I had another accident on a roundabout on the infamous North Circular Road when a truck all but drove over the bonnet! One front wheel was almost flat on the road. Next morning I fronted up to the transport department again to tell them where the now undriveable car was to be found. This time there was a raised eyebrow! I got the distinct impression that I was considered to be a high-risk driver as they assigned me my third temporary car within a few weeks. My brand-new car eventually arrived and I managed not to have any further incidents.

I spoke to a meeting of the whole department after a few weeks and after I had interviewed them all individually. I laid it on the line that we were currently losing money, we had to change this around, we had to produce good systems that the clients wanted and at the right price. It all sounded easy, but there was a huge amount of work to do to put things right. I laid down a few objectives and declared the new organisation – I promoted new team leaders to front up to the various businesses. This seemed an obvious thing to do but hadn't been done before and was one of the reasons that I.T. – business communication had been dreadful. I told them that I couldn't do everything on my own – I would need their expertise. I was going to demand high quality work and it wouldn't be easy going. Anyone who didn't want to work hard should consider leaving, I wasn't going to try to persuade anyone to stay. I think that it went down well.

One of the first things to do was to run an advertising campaign to recruit the missing staff. I was told that it was nigh impossible to recruit to a dirty messy factory in Ilford, a rather grubby east London suburb when I.T. people could get very well-paid jobs in the city in much more salubrious surroundings. Of course, my staff were pressing for higher salaries, saying that this was the only solution! After some thought I didn't advertise in the usual places, but restricted the adverts to the local press. I aimed them at people who lived in Ilford and Romford and travelled to the city to their jobs. I focused on lifestyle and informed them that just a few minutes away from home they had an opportunity for an interesting job, saving much travel time and money. It worked! I recruited a good number of well-qualified people and none of them left. I did however stuff up severely on two people. Soon after one chap started he began to come in late, wearing dark glasses. Then he didn't come in until lunchtime. Then not at all. I found out that he had used his I.D. card to gain entry to other Plessey sites so I put out an alert for him. He was apprehended at a site when he tried to obtain cash there from the cashier's office. He was clearly an alcoholic and the gaps in his c.v. were now apparent for what they were. The personnel department had let me down in not following up the references and he had lied to cover the gaps. As well as wasting all the recruitment time I had the lost opportunity costs to cover. I compounded my error by recruiting a second alcoholic soon afterwards! I was informed one morning that the client meeting he was attending was going badly. I went to the meeting room. He was slumped across the table slurring badly and mouthing drivel. I yanked him out to my office. "You're drunk," I said, "and that's an offence on a site with machinery, apart from anything else. Go home, and come back tomorrow morning sober and we'll discuss this." I had him escorted to the police gate. That was the last I saw of him. I received a letter from his solicitor informing me that I was to be sued for wrongful dismissal, on the grounds that I was drunk at the time! The company said that they would back me in fighting this, and the case was dropped soon afterwards. The personnel department informed me that I held the record for recruitment of alcoholics! These two were both very well qualified, with some good experience, and when sober could produce good work. I had even helped one of them to get a flat in Ilford, loaning him the deposit from department funds. We were able to retrieve some of that from the monies owed to him. I'm still unsure whether I would recognise another alcoholic – they were such good liars!

Another would-be recruit was put my way from Joe. "Have a look at this chap, Derrick," he said, "he seems just right for you." I was uneasy at the interview, his answers were a bit glib, but I was unable to put my finger on the issue. I thought about it over the weekend and on Monday I telephoned the university that had issued his degree. "He attended the first year," I was told, "but failed the examination. He was invited back to repeat the year but he didn't return." I told Joe that we had a fraud here, and he was impressed that I had established this. In the "no, thank you" letter I was very tempted to tell the chap that I had discovered his fraud, but the personnel department advised just to turn him down without mentioning this. It made me wonder if this was a common situation. While writing this in Melbourne I read that a very senior executive has been fired from a communications company for just the same reason!

I soon noticed that one project was losing lots of money. It was an early 'on-line' system development (most were batch systems then) and was staffed by contractors. I established that the users were constantly changing the requirements and that the documentation was non-existent. Controlling this sort of project is impossible for all sorts of reasons including getting the customer to pay the appropriate price! I decided to cancel the project, sack the contractors and replace the system with a modified version of a current system, so that we could control the development properly. Some egos and kudos were hanging on this development so I had to argue my case up the management chain to the I.T.

director to obtain his agreement. I couldn't understand why people were showing surprise that I was going to argue this case with the director. I found out that he was a very sharp character who would not hesitate to tear you apart if he thought you were wrong. He was tall, athletic and very bright. His nickname was "golden bollocks". He gave me no trouble but reluctantly agreed to my proposal, and I then had to mollify the customer.

At first it seemed that all the staff were very busy, while we had a big list of work not started. And yet our invoicing didn't reflect the work done. When I looked into it I found that much of the work was unauthorised – it was 'tinkering at the edges' and 'gold-plating' – not work requested by the business. Computer staff often tend to be perfectionists – it goes with the job, certainly for programmers. A full stop in the wrong place in the computer code could be disaster, but many programmers would spend days modifying the code to make it run faster. They would often not see the wood for the trees and not realise that they were only saving a few microseconds. I insisted that no work would be performed without written authorisation, signed by the client business manager – thus allowing us to invoice. I immediately had a stream of staff without work. We were able to start the jobs that the businesses really did want – and would pay for.

I had a bookkeeper, who used various computer systems to produce detailed monthly financial results and summaries. I had to pay for everything – my own car leasing costs, maintenance, fuel, heating, lighting… and salaries. I had to quickly become accustomed to understanding the reports and to make decisions on where to spend money, push for more client work, or to recruit more people to meet demand. Once a year I had to produce a year's budget, against which each month's results would be compared. It was a horrible job, taking me several weeks, wrestling with projections. "If I get this much work, how many staff will I need, what type, at what cost? What would the overheads be? How much useful time do we get from each member of staff? What is the average sick time?" ... and so on. I tried to closet myself away to have a quiet, undisturbed time, but that rarely worked for long. The first budget contained a lot of guesswork – I optimistically surmised that there was a reasonable opportunity for a lot more work, and I budgeted on this basis. It turned out to be valid – the turnover – and profit jumped. I managed though, on a month-by-month basis, without paying much attention to the budget. I took decisions based on what was happening, rather than what the budget said. That seemed to me to be most sensible! I rarely used the accountants – they seemed to live in another world to the one that I inhabited.

Some of the work being done in my department was extremely sloppy. I made it clear that I wouldn't tolerate this and I put in some quality assurance processes. All reports emanating from the department had to bear my signature, and on day one I had been asked to sign a weighty document. I had said that I'd have a look at it, much to the surprise of the author, who had been expecting me to sign it there and then. Next morning I gave him the document back, unsigned. I pointed out that the English was awful, with grammatical errors and poor spelling. One page had scarcely one full stop! "What credibility will our customers have in us for our technical ability if they can easily find fault with our English?" I said. He went back to his desk looking dazed. I asked my secretary to buy a good dictionary and a thesaurus and encouraged all the staff to use them.

The Plessey businesses that I serviced were each relatively free to pursue what business they wanted, as long as they stuck to the rules and made a profit. I had to front up to the various managing directors and senior managers to service their I.T. requirements, make recommendations and so on. This meant that I was often running around to meetings all over the country, and of course I had to chair many myself. Once or twice I hired the company helicopter to get to meetings at short notice, once taking off from a football pitch at Dagenham and another time I had a meeting at Millbank on the Thames at the Plessey directors' offices, where the chopper took off and seemed to climb the Battersea power station chimney! This was expensive of course and all had to come from my department budget, so it wasn't something I did every day. We also used a fixed-wing aircraft to get to Newcastle on one occasion when there were several of us to get there. Plessey had rules for this. "Staff in categories X,Y and Z are entitled to first class travel but only if..." sort of thing. These were the rules I ignored and I just organised cars, trains etc for my staff that seemed most sensible for the situation. If I was challenged I would usually apologise – it was the quickest solution. I found it was better to do something and apologise for it afterwards. This worked well as long as we had success. And we did. Working with my new team leaders we turned around morale, we started to produce what the clients wanted and we lifted the department from the bottom of the various league tables that were published each week to a good position. Success breeds success. People love to succeed at their job, and when they do, and when they get recognition, they want to do more. So we began to have fun – and make a profit. But some of that took a year or more to achieve. I had been given six months to make a difference and I passed that milestone. I got used to making quick decisions because I had little time for pondering. I've met many people who kept putting off a decision and seen the poor results. Sometimes you can take more time, if nothing costly or disastrous is going to take place in the meantime, and if you can use the time to collect more information about the issue that will help the decision-making process. But sometimes the decision just has to be made. If it's the wrong one, then you'll find out soon enough, and you can put it right.

The new teams I had created were working, making it much easier to foster a team spirit and to improve communications, both within our department and with our customers. When I spent money on the department office – we were in a factory-type building that became very hot in summer, having a glass roof – I delegated each manager to reorganise his own team as they wished, and gave them a small budget for plant pots and furniture. I had the glass fitted with reflective material that made a great improvement. These were small things but they made a difference. I also made a point of walking around the department each day that I was in the office. It was easy to forget to do this, but I knew how important it was to be visible to staff and to stop and chat to them. The staff responded well and produced good results. I tried to ensure that when there was a good result that the senior client manager would thank us for doing a good job, and that the thanks went directly to the team involved. Sometimes a quiet word was necessary, for I was aware that an 'unsolicited' but genuine word or two of praise, especially if made in public, does a power of good for the morale of the people involved. It works better than a pay rise! So many managers seem quick to criticise but slow to praise. It costs nothing, and is so effective – but it must be genuine and deserved.

Each year Plessey took part in the 'milk run' – where teams of Plessey managers would visit UK universities each year to recruit graduates. I was asked to join one of these teams and we toured the country for a week or so, spending a day at each of our target university sites. We would be recruiting for a range of vacancies and for a range of Plessey sites. The interviews were one-on-one, and were planned for one every half-hour, with twenty minutes for the interview and ten minutes to write up the results. It was exhausting, but quite fun. It was interesting to observe how the applicants presented themselves – some were well-prepared and had done some homework on the company, some hadn't a clue who we were or what we did. Others turned up looking as if they had just got out of bed. We decided who we would like to seriously consider and they would be invited to attend a Plessey site for a full-day interview. I would manage these full-day interviews for the Ilford I.T. candidates, and my team leaders would be involved in this. I would provide lunch for them and the first year I hit one of the Plessey rule-book issues. The company was very hierarchical and had many military-style customs and practices. I was classed as 'executive manager' and as well as qualifying to use the 'executive mess' I could invite my visitors to the 'executive visitors' mess'. I duly booked in my fifteen candidates and my team leaders. A few days after the lunch I was taken aside by the executive mess chairman and gently informed that our visitors' mess (there were others for other ranks) was only for visitors of managerial ranking!

Plessey was managed by the ex-army Clarke brothers who were reputed to be a tough duo. Indeed Plessey had a 'hire and fire' reputation. Michael Clarke made a lightning visit to the Ilford factory one day and I got word that he was on his way to my end of the woods. He came through very quickly and barked a few questions. Then he got to the punch room. Now I wasn't in charge of either the computer operations or the punch room with its twenty five data preparation girls, although I was very involved in various ways as my clients would often make negative comments about this part of the business and I took some flak. It was frustrating to me that the computer operations had an entirely different management structure. Michael Clarke saw the lack of activity and apparent chaos in the punch room and asked me what the hell was going on? I said that I didn't have responsibility for this; more's the pity. He looked pretty angry. A short time later I was informed that I was now in charge of the computer operations and the punch room! Again there was a huge amount of work to do. The operations department was a mess and the punch room was worse. It was bleeding money at a rate of knots. When I looked at the numbers I could see why. The absentee rate was twenty per cent. On average each girl took one day each week off – sick! It was clear that they had some racket going on and took turns. I read the riot act to them and made it clear that I knew what was going on and it was going to stop – one way or another. I had to be a little careful as the place was heavily unionised and there were rumblings. At first nothing happened – the absentee rate continued. I found out later that they thought I was a pussy cat. I then laid out the rules. They had to call in and speak to me, or if I wasn't there, to the supervisor if they were not coming in. The situation slowly improved, but not before I had a number of the girls in the office making their excuses. One of them tried to embarrass me with her tales of 'girl trouble'. I said that we could help and that I would arrange for the company doctor to see her. She quickly got better. I also realised that their job was not the most interesting in the world and tried to devise ways of improving things. After consulting with them we organised the work differently and installed a little competition between them. We measured their production rates and published them. We improved to the point where the section made a profit and I wasn't spending too much time on it.

Part of my job was to sell our services and one managing director refused to accept my offer to build him a system, saying that he had two engineers who could write computer code and he would go ahead with them (managers often think that systems are just about computer code). I pointed out that this would make him vulnerable, as he was relying on them alone. They might well be excellent at writing embedded code for hardware, I said, but they probably did not have the disciplines and skills required to agree and define a big commercial system and document it, quite apart from designing, testing and implementation issues. He responded that they didn't need to define the requirements, because "they all knew what was wanted". I've heard this so many times – and also heard the subsequent arguments when, after spending hundreds of thousands of pounds and the system has failed – "but I thought that it would do…" and "… but surely you realised… " and "… why doesn't it do…". Without a formalised, agreed and signed off document there is no hope of success. My points were all politely ignored, so I wished him well with the project. At the monthly meetings I had with him, I would casually enquire how it was going and was always told that it was on track. After six months the answer changed. Both engineers had resigned, the system was "almost finished" and would I please pick it up and complete it? I replied that people didn't usually resign when the successful end of a project was in sight, so I suspected that it might not be as he had been told. And both resigning together? Nevertheless I would put someone onto it and report back.

What I had to report later was that all we had was a mess of computer code, no plans, no documentation of any sort existed and no way that we could fathom out what was going on. And there was no agreement on what the system was supposed to do! The only way forward was to start from scratch. Of course he wasn't pleased to be told this but quickly agreed to sign up with me for the project. He had lost a lot of money and six months opportunity costs. This was another of those times when I found it very frustrating to be unable to convince someone in authority of the best way forward. Very often senior people become convinced of their infallibility in decision-making and ignore the best advice available.

There is another – hidden – issue here. Software systems, by their nature, are expensive complex beasts. But you can't see them. (You can see screens, documents, but they are just ouputs.) There is nothing to grab hold of, nothing recognisable to look at, so a manager who knows little about software finds it difficult if not impossible to understand why a system that he or she wants might take twelve months to build and cost one million dollars. If that person was contracting to buy a house or an office block there would be plans, recognisable artist impressions, even a 3-D model to look at and criticise and sufficient comprehensible information would be available so that a lay person could grasp the fact that this building could not be available in, say, three months. But the proposed software system is not available to look at, to poke and prod. So the impatient manager who is used to getting their way may demand completion within three months. For similar reasons it's difficult to measure true progress – one can see a half-built house but there's nothing to look at when five hundred thousand dollars has been spent after six months of work on software. These issues sometimes make a meaningful dialogue all but impossible with certain management types who try to thump desks and bully staff into achieving the impossible. There are ways around the issues but it takes more than a five minute conversation. Sometimes that's all you have.

I tried to sell an online stock control system to one business that sold aircraft spare parts. The old system that they were using was pretty useless but again I couldn't persuade them to take the plunge. One day an American air force colonel, one of the business' best customers, arrived on an inspection. He wanted to know why the spare parts never seemed to be in stock in the USA for his helicopters. He asked to see the system that was used to manage the supply. The M.D. of the business was reluctant, but he had no choice. The colonel's face was a picture when he saw the manual card file that was used to keep track of the one hundred thousand parts on record. The cards were stored in metal boxes on long wooden benches, with staff wandering among them taking out cards and trying to maintain them. It wasn't far removed from Victorian times. You can perhaps imagine the problems that arose! "This is ancient history," he said. "If you can't quickly do better than this you will lose all of our business." His actual words were rather more colourful. The next day the M.D. signed the order for the system. As part of the preparatory work we checked out the parts file and established that approximately half were 'dead' – never called for. We even found parts for the Spitfire there! The trouble was to identify these and then weed them out before the remainder were converted to a computer file – a costly, time-consuming business as much effort had to go into writing software to edit, check and convert the records. The old stock was sold off to some enterprising chap who tripled the prices and made his business selling to those customers who were still trying to maintain ancient aircraft. Apparently aircraft are sold in much the way that old cars are sold and many old aircraft end up crop spraying or being used for passengers by a third world country. Not only did we install the system but we installed an online link to the USA that ensured that the records were up-to-date. One of my analysts was excellent at this kind of system and was terrific at producing them at short notice. This was still in the Seventies, so these things were not yet commonplace. The M.D. was delighted with the results as was his client in the USA.

One thing that one gets used to as a manager is listening to staff who have problems, be they personal or business-related. One of the operations staff came to see me one evening. She worked on the late shift, keeping the computers running and the links to the computer centre open. She had an issue relating to her treatment under the previous regime, pay and conditions, she was unhappy and wanted something done about it. I listened and said that I would look into it. I did so and was able to obtain a satisfactory outcome for her. Shortly after this she turned up again with another issue, relating to the past. Again I listened and again I managed to fix it. This all took up time and I was very time-short. There was much to do and I rarely had time to just think. I always worked until at least 6 p.m. and it was in this quiet time that she would come to see me. The next time she arrived with yet another issue from the past I'd had enough. "I'm sorry," I said, "you have to accept the past now and get on with the future. I am not able to do any more for you." I'm not sure whether she really felt aggrieved over past issues or whether she just wanted attention.

Plessey was a big supplier to the military – wireless and communication systems, torpedo manufacture, instrumentation and so on and so each site was surrounded by high fencing, with company police always on duty at the gates. One day two burly fellows in blue suits came to see me. "We're from security," they said. "We are going to check out your department." After a while they returned. "This isn't good enough," they said. "You have desks overflowing with computer code and paperwork, it's left there at lunchtimes, even overnight. Anyone could take copies of anything." "And your files," the other said, looking at my desk, "aren't much better." I defended our position, saying that most of the systems work that we did was humdrum, accounting and manufacturing systems, stock control, normal commercial work. And even if someone got hold of any of this, it wouldn't make any sense. "I can't understand much of it myself," I said. "That's not the point," he said, which I thought was precisely the point. "Suppose that one of your staff was a bit disgruntled," he said. "He could take a copy of the documents, slip it to someone in the pub, and return, and you would be none the wiser." True, I thought, but a most unlikely scenario. They got a bit cross with me as I was clearly treating the situation too lightly. "I suppose you'll be telling me that I need a wall safe behind the picture next?" I said. They looked at me with part-closed eyes, their looks and voices becoming colder. It was made clear that I had to ensure that all paperwork was put away, that my files were locked, with a large iron bar locked into position down the front of the filing cabinet, and the key – yes – in a wall safe. This was way over the top in my view but I had no choice; these chaps reported to the M.D. They gave me a demonstration of spying, to win me over. A terminal was set up – the first computer terminals were now in use – and in another room some hundred metres away the signals from the computer screen were picked up by some small radio receiver and shown on the screen of the spying terminal, inside a suitcase. "If we can do it, so can they," I was told. I didn't know who 'they' were but I was impressed. This was James Bond stuff, but I didn't see how this was of concern to us. None of our work would have been of interest to any foreign country as far as I could see. Who wants data from the payroll system or the accounts payable? And so what if we had 2,035 bolts in stock that fitted a particular torpedo? I wasn't convinced, and the security chaps said that people like me were part of the problem. Maybe I was, but I reckon that they had seen too much of 007, but I had to comply with their requests.

I charged out the staff at different rates depending on their seniority and the job. A senior project manager would earn a high rate for a consulting job for example. Joe – the mafia-type chap, who was actually quite easy to get along with – was looking at my monthly financial report. "I notice," he said, "that you do not do any consulting work yourself. Your predecessor did, and earned good revenue. Why don't you?"

"Do you want me to be a consultant and stuff up the department like my predecessor, or manage the department properly and make a success of it?" I asked. He looked at me and his face slowly cracked into what was the nearest he ever got to a smile. He, though an experienced manager, had fallen into the trap of confusing *doing* the job with *managing* the job. I'd seen many technical people who, on promotion to a managerial position still try to work at the detailed level with disastrous results. I wasn't going to do that.

My three years here were fun, although the first year in particular was hard work, and I would often be driving home around the North Circular Road quite late. I had a hand-held dictation machine and I would dictate as I drove. I was stopped by a police motor-bike rider one evening who warned me for speeding – and for dictating! Now of course it is illegal to use a mobile phone while driving unless it's hands free. And quite right too. My secretary, Joan, was a lovely mature lady, terribly efficient, good at shorthand and thoroughly professional. I was able to rely on her to look after the office, handle messages and sometimes stressful situations without becoming at all ruffled. Her typing was first class – and she could spell! I was lucky with her. She was my secretary the whole time I was there and I also took on an assistant for her as the work load expanded. This was before word processing came along so everything was typed from shorthand, hand-written notes or dictaphone tapes.

I was enjoying the job, and I had succeeded in doing what I'd been recruited to do – the business now turned a good profit, morale was good, staff turnover was almost non-existent, and the customers were generally happy with the results. I felt that I was using my accumulated skills and experience to the best advantage, I felt confident and rewarded – and I was still learning! One day in 1979 I had a phone call from Grahame, a chap I had recruited while at Sainsbury's. He had left and returned to his previous job – as a partner/trainer in a tiny I.T. training company. Business was looking up, he said. Would I like to join them? I had this hankering for running my own business, and although I was doing that in a sense, it was still Plessey. I was a senior manager with some seventy staff and the trappings, authority and prestige that a large company provided. How much further would I go? Would it be worth it? Would I have better prospects and satisfaction in a tiny business of four people? I would become a presenter of training courses with no back-up staff. But I knew that I could speak reasonably well, people seemed to listen when I spoke at meetings, and I could hold an audience when I gave a presentation – and I enjoyed it. I ruminated on this for a month or two. Then I rang Grahame. "I'm in," I said.

Some months after I had left I received a call from the recruitment consultant who had put me forward to Plessey. "I had a call from Joe (the mafia-like manager) the other day," he said. "He wanted me to recruit another manager." 'Get me another Derrick Brown,' he'd said. "I told him that there was only one Derrick Brown, and that he'd just let him walk away." I was flattered at this, naturally, and hoped that I'd made the right decision! On reflection now although the experiences would prove to be useful in many ways my managerial skills were never stretched any further from my Plessey days. Instead my personal speaking and influencing skills would be put to the test.

CHAPTER 7
A Career Change

Keith London Associates was based in the small Hertfordshire village of Welwyn. Apart from Keith himself, there was Colin and Grahame – and now me. Colin's wife Liz did the books and administration. That was the company, a bit different from Sainsbury and Plessey. The office was in a loft above the garage at Colin's house at first, but we quickly rented an office in a small office block in Welwyn Garden City. I wondered for a time whether I had made the right decision – I missed the power and the glory! The business was running short training courses for computer people – mainly the systems analysts but some others as well. The big companies all had staff to train, the computer industry was growing fast and the business had begun to earn a reputation for good quality training. I tagged along to experience some courses. They ran for between one and five days, were either public courses or in-house (for one company) and many of them were residential. We would hire a hotel, often tucked away in the country, and stay there for up to a week with the delegates. I soon realised that the fees being charged were more than reasonable – they were far too cheap! It took a little persuasion but we increased them and this did not affect business as my colleagues had thought. The founder, Keith, had interviewed me at 'The Flask' public house in north London. He was a character and a most entertaining presenter, I was told, although I never did see him presenting. Shortly after I joined we received a telex (no emails then) from him. We had some business in Australia and he was there running a number of courses. FALLEN IN LOVE STOP NOT RETURNING TO UK STOP GOOD LUCK STOP KEITH. He never returned, he left the business but we retained his name. He had had several wives and partners apparently but this particular lady managed to pin him down.

So here I was now speaking for a living. I had always quite enjoyed public speaking, but this time I had to run a training course and keep the students' attention for seven hours a day for up to five days. And as their companies were paying good money the students' expectations were sometimes very high. The course material was not new to me, I knew from experience what their problems were and so I found that I was able to relate the material to them. This was essential, as any course of a few days in length can be a dreary business unless the presenter livens it up. It was not uncommon then to find a presenter who just read from the manual! I was able to relate anecdotes and stories that backed up the material. Over time I worked up a number of funny stories (mostly true) and I would fool around a little and use a Basil Fawlty-like character that went down well. A student told me that I was a cross between John Cleese and Noel Edmonds (a high rating disc jockey of the era). I came to enjoy this and it was easy to extend, I found, as I had a wealth of characters and experiences to call upon. I had to be careful that I didn't overplay these aspects, as I had to get through the publicised course material. It clearly worked, as I once achieved our highest possible student assessments of top scores in all topics (never done before or after). This was almost impossible, as they were asked to score the venue, the food, the course manual and materials as well as the presentation, objectives... To get all sixteen students to score a five for all topics was unknown. Clearly, if they enjoyed the course sufficiently, they tended to score the venue, etc more highly than they otherwise would do. We would send a summary of the scores to the students' companies so this had an effect on future bookings. As I enjoyed this aspect so much it was not difficult for me to continue along these lines. I had to learn to adjust the presentation to the culture – the same jokes and stories (usually against myself) went down differently in say Dubai to Milan. A most important part of every course was the practical work, where we put the students in a situation, usually in teams, and they had to work on the issues and present their solutions. The course manuals that we provided were at first nothing more than copies of the visual aids that we used, but we soon began to add text, and they became more like text books. Over time they took on a most professional look.

At first I missed some of the rough and tumble of the 'real world' of Plessey, managing a business and lots of people. Here the only money we made was from the business that we sold and then delivered. And we had to do everything, or be involved in it. We began to build the business. My colleagues, who became my partners after they offered me a partnership, were initially reluctant to grow, foreseeing growth issues, including management. I argued strongly for growth, otherwise we were never going to make much money. So we took on more presenters, hired an office administrator, bought a word processor (a purpose-built computer that only handled the one application) and expanded the office accommodation. Then we took on a salesman. It was all quite fun, and although we were fairly conservative the money came in and we had no debts. We all had company cars, bought for cash. After a few years the offices were bursting at the seams so we moved out and took on an old printer's premises at Welwyn which was a total wreck. It needed a new floor, new roof, plumbing, wiring ,,, so we had the whole place refurbished. It had some woodland at the back so we cleared a little to make a seating area. I'm not sure that we ever found our way to the rear fence. One day a small deer appeared and looked at us before bounding away. The office took shape and we moved in. Eventually we had half-a-dozen administration and word processing staff, a print manager (for the manuals), a team of around fifteen presenters, some thirty staff in all and about twenty company cars, all fully paid for.

I was working madly, often running courses back-to-back. I'd often leave home on Sunday evening, drive to the hotel we were using then get home on Friday evening. At first we used a hotel near Gatwick airport for most of our public courses and I soon knew the routes blindfolded. Then we began to run more courses overseas – Australia at first, where we had an agent, but then companies like Burroughs Computers, Hewlett Packard or Price Waterhouse would ask us to run courses for their overseas sites, so we would trot off to Milan or Munich or wherever. Then Colin found a contact in Dubai and we began to find opportunities there. We employed an English woman as a part-time agent for us and she became a great asset as she had many contacts with the locals. We ran many courses there over the years – I must have gone there at least half-a-dozen times myself. We branched into Saudi Arabia, Oman, Kuwait, Abu Dhabi and Bahrain and I ran courses in all these places. We would combine courses into one trip so that I might go to Dubai, then onto Oman, then Bahrain, or sometimes we'd combine a course there on the way to or from Australia. On my first visit to Dubai I went for a walk as soon as I had unpacked. Dubai then was just growing up from what had been a small dusty town and I took a ferry across the river. It was a dhow, steered and rowed by two men, and packed with people – all men, I noticed. I wandered about, visited the gold souk, and began to feel uncomfortable. I then noticed that many of the men were walking about in couples, holding hands. Then I realised that there were no women to be seen, not one. Welcome to the Arab world of men! There were no females working at the hotel, no female chambermaids, not a woman anywhere. This made me feel very uneasy. In the restaurant there was a separate section at the back, with a brick wall separating the section where women and children could sit, but not be seen. I developed a foot infection once in Dubai and went to a doctor's surgery. Apart from the receptionist, there was just one woman in the waiting room, all dressed up in her hijab, top to toe. As soon as I appeared she asked the receptionist for a private room. She wasn't even allowed to sit in a room with a strange man! I felt like a leper! Women were not allowed out by themselves, they could not drive cars, they had to be completely covered. I raised this cautiously with one of my students on my first course. He was very friendly and most polite (they all were) and knowing that he was married I asked him what was the difficulty in allowing his wife to drive the car, for example? He replied that he understood my difficulty and that he had no problem with it at all. "But" he said "if you ask me 'Would I be the first person to allow my wife to drive?' I would have to say 'no'."

One course I ran for an oil company in Saudi ran into trouble. There were two chaps in the Arab dress who just sat there talking to each other while I was speaking. The normal way of handling this sort of audience problem is to stop, and just let them realise that you are waiting for them to shut up. This succeeds nearly one hundred per cent of the time. It didn't work with these two. I walked down to them and asked quietly if there was a problem, and could I help? They looked at me and said nothing. I repeated the question. No response. Then I asked them if I was making myself clear. Did they understand me? No response, just a rather arrogant look. I gave up and returned to business. At coffee break I tried again without any response. I spoke to the administrator and suggested that as they didn't seem to want to be there that they should be removed. He explained gently that they had senior positions with the company, that they were both completely useless, but that was part of the deal that the company had to put up with if they wanted to be in Saudi Arabia. They attended the course just for appearances. And they could easily make trouble for me! After that I ignored them, they ignored me and I didn't expect them to do anything. And they didn't. I learned that to do business in Saudi a company must have Saudi directors on the board, and employ a certain percentage of Saudi nationals. No company or foreigner may own property there either. In these and many other ways Saudi Arabia prospers!

My first trip to Saudi included a talk to a room full of senior managers. It was to introduce computing concepts to them and to provide a background of what was happening in the computer world. Talking about data input I showed them a cartoon of a data preparation operation. A data prep operator sat at machine, smiling brightly. She wore a mini skirt and was showing a bit of leg, contemporary with the times. At the break I was taken aside by my contact. "It isn't done to show a picture like that" he said. "It's OK with us, we're mature, but don't show it elsewhere!" I was embarrassed and a bit cross that I hadn't been forewarned of this by my partners who had already done work here. I must have been a little naive as well. I also showed a visual aid of a cheque, to show the magnetic ink characters that are read by machine. The cheque was made out to Marks and Spencer! I had a bit to learn about doing business in Saudi! At the airport I looked at the newsstand. A poor array of books were available. A cheap detective novel caught my eye – it had a lurid picture on the cover of a man with a gun and a woman at his side. Her dress showed some flesh at neck and arms. This had all been crayoned out with black crayon! I imagined the censor at his job, slobbering over his work, thoroughly depraved. Then I bought a newspaper. There was an article on the famous Rodin sculpture, 'The Kiss' with the entwined couple, quite naked. They had been blacked out completely, except for their heads! I wondered how many people were employed at this entirely useless function, blacking out pictures in every copy of books, magazines and newspapers so that the population would not become depraved.

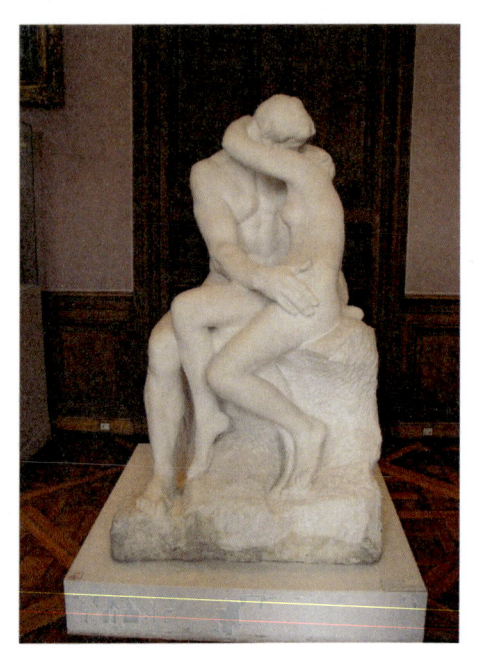

What the Saudis weren't allowed to see!

Another time I was invited by a computer manager to stay at his flat in Riyadh. Staying by oneself in hotels quickly becomes boring so I agreed. We had a Saudi weekend to cover. On Friday, their Sabbath, he asked me what I wanted to do. "What do you do usually?" I asked. "Shopping," came the reply. We drove into the city centre of Riyadh. On the way we were stopped by pedestrian traffic in a busy thoroughfare. "Where are all these people going" I asked. "To the stadium," he said. "Today there's an execution and several amputations to watch." I declined the offer to attend. In fact we had to move away quickly because anyone in the vicinity would be encouraged by a little beating from the religious police to attend. They liked these punishments to be well-attended. Back at his flat after an exhilarating shopping trip at the supermarket he suddenly whispered to me, "Do you want a drink?" I almost had to read his lips. "Yes," I whispered back, puzzled. He put a finger to his lips and motioned me to follow him. He unlocked a door to a bedroom, unlocked the wardrobe and there was a small still! He carefully poured out two glasses of beer. Smiling, he mouthed the word "Cheers!" It was terrible beer. Apparently he was convinced that his flat was bugged. He had returned from work one day to find two men who were 'repairing' his phone. "But I haven't reported it broken," he said. "There's nothing wrong with it". "Yes there is," said the men, then left. Neighbours were encouraged to report any wrongdoings, especially by foreigners. Drinking beer privately, although frowned upon, was OK for foreigners but brewing was definitely not. A public whipping and prison was the punishment for that. He and his wife had come out from the UK to serve out a contract, but his wife returned home after a few months. I didn't blame her, I thought it was an awful place, quite apart from the lack of good beer.

The British inventor Clive Sinclair brought out his home computer the ZX81 in 1981. We bought a couple (they sold for £70) and played around with them. They had very little memory, no operating system and no software apart from games so were of limited use. Someone had the bright idea of putting them on a stand that we had agreed to take at an exhibition in Dubai. Training is not an easy product to exhibit, but we tried, and had training books and a display to show what we did. We thought that the ZX81 would be a draw card – and it was. The locals just pulled out wads of notes and cleared our stand within minutes! We telexed back to the UK office "Send more!" but the Sinclair organisation was hopeless and wasn't able to respond at all.

In Dubai we used a training room in the Dubai Tower. It was one of the few tall buildings then, on the edge of the desert. It bristled with antennae, as the CIA were rumoured to have a floor there. That explained the armed guards on the ground floor! One floor did not appear on the lift controls, so it was obvious that was it. They must have had their own lifts. Once I was staying there when a member of the royal family was getting married. The desert became a city of tents, with hundreds of camp fires and magnificently dressed Bedouin warriors, with jewelled knifes at their belts. Some carried ancient muskets, while the young bucks drove around in their four-wheeled drives firing off AK47s. Phil, one of my colleagues and myself wandered around the camp fires. There was much singing and dancing going on and it was all rather fun. We took a few photographs, very discreetly, as in Oman I'd had a very bad reception when I took some photographs of women washing clothes at a public washing trough.

Dubai was a better place to stay as it had rather more relaxed laws. Arabs from Bahrain would drive over the border at weekends to let down their hair. In Dubai we struck a deal with one hotel that we used for training whereby the delegates could stay there at a special rate. I enquired on one course if many were staying there. None of the people from outside Dubai were staying. "Why not?" I asked. "It's better at the one next door" they replied. I established that the one next door had hot and cold running women in all bedrooms!

We sometimes used training videos and the best were ones from Video Arts, the company in which John Cleese had a major role. I was taking one into Saudi Arabia once when it was found by the customs official in my suitcase. "What's this?" he screamed at me. "Pornography? How many more do you have?" I told him it was not pornography, it was the only one and that it was a training film. He rummaged through the suitcase, then took the video saying that they would have to view it, to see whether it contained pornographic material. I was not going to get it back until the next day. The issue was that I was about to get a connecting flight, and I needed it the next day! When I protested a soldier stepped forward, put a gun in my stomach and said "Come with me!" They took my passport and locked me into a small, windowless room. It's a bit frightening to have your passport taken away, especially in a place like Saudi Arabia but I was more angry than frightened. I was there for thirty minutes and my plane was scheduled for take-off. I banged on the door and it was eventually opened. I was bundled into a mini-bus and taken onto the tarmac to my plane. The passengers all glared at me, as they had been told that they were waiting for a late passenger! At the client's company the next day I explained why I would be unable to show the video. Without murmur the manager detailed a minion to drive the several hundred kilometres across the desert to the airport to collect the video. I wondered whether the censors would have been salivating at the opening shot of the video – it showed John Cleese in bed with his screen wife. They must have been disappointed with the rest of it!

Another incident occurred concerning a video. I had shown one at the Dubai Tower to my course and one delegate hung back in the evening. "I'd like to have a copy of that," he said, "We can't buy them here." I knew that Video Arts would not sell their products in the Middle East because they knew that pirate copies would be on the market in no time. "Sorry," I said. "I can't allow you to copy it." He looked surprised. "I'll make it worth your while," he said, producing a bundle of notes from inside his flowing robes. Again I shook my head. He was more surprised. "Look," he said, "just leave the video on your desk and it will be there in the morning. You won't have done anything wrong." "Sorry," I said "that just isn't on." He looked at me as if I were quite mad. He really didn't understand.

I landed at one airport which was located all by itself in the desert and I went to get a taxi to a city hotel. A large noticeboard declared the fares which meant that I wouldn't have to go through the usual haggling process. However the first taxi driver quoted me a fare that bore no relation to that quoted on the noticeboard. I checked with the next taxi... and the next, but they all had the same story. I could see that this was a racket so I approached a policeman standing close by, with his AK 47 cradled in his arms, plus a sidearm. I asked him if the fares quoted on the official board were current, and was this what I should pay? He made no reply. I asked him again. He looked away and was silent. He was probably paid some baksheesh by them, I thought. I had no alternative but to agree the fare asked. I got into the cab with its torn and worn seats with my luggage and training materials. A few minutes into the drive my driver stopped the cab and turned off the engine. It was over one hundred degrees and there was nothing but featureless desert to be seen in all directions. He turned to me, his unshaven features and swarthy appearance suddenly looking menacing. "The fare is," he said, and quoted a fare three times that which we'd agreed a few moments before. As he spoke I was looking at his mouth which displayed mostly empty sockets with a few teeth looking like icebergs in a dark sea. I saw red. I didn't think of my precarious position at all, I was just furious. "You bloody crook," I yelled at him, "you'll take me to the hotel for the sum we agreed, or there'll be real trouble!" He turned round, restarted the motor and drove on as if nothing had happened. At the hotel I paid him the agreed cash, still furious with him, and he almost smiled. I thought later what might have happened had he put me out...

These incidents and others like them had given me a rather negative view of Arab culture and business practice. Now I have a slightly better understanding, having done more reading. It appears that by tradition a man's *honour* is of the highest importance. This is a complicated business and affects all manner of interactions. The avoidance of *disgrace* is paramount. A man is under obligation to feed and clothe his family, for example, and to make the best of any opportunity that comes his way to make money and improvements to his immediate and extended family. Thus a man who has a position of authority or responsibility also has an unspoken obligation to use that position to better his relatives in whatever way he can. Nepotism, of course, is frowned upon if not being actually criminal in much of the world, but it was not so in the Arab world. Similarly one would be expected to wring the most out of any business negotiation, even though others might call it a dubious practice. With this in mind I might look upon some of my incidents just a little more generously. I read that these traditions have changed or are changing as the countries concerned develop their economies and trade with the Western world much more, but the Arab and Western cultures remain a long way apart.

Hong Kong became a popular business area with us. We would stay at a rather swank hotel, The Excelsior, and sometimes ran the courses there. Having an evening drink on the top floor bar with the lights of Hong Kong spread out below remains one of my more vivid memories. We found an agent there and did some good business until the agent sunk all his money into software development and went broke! He owed us our share of the income from several visits (about £10,000), but we never got a cent despite his many promises. It was there that I went for a massage once as soon as I arrived at my hotel. I changed and stepped into the lift to go down to the basement pool and gym. Standing in the lift I became aware that my jeans felt greasy... and loose around the waist... and short in the leg – they weren't mine! My brand-new Levis had been exchanged for this old pair by some enterprising baggage handler! Annoyed and also somewhat amused by this I went to the pool area and saw that they had a massage parlour there. "Can't miss this," I thought so after my swim I went to the desk and paid for a session. I had in mind a honey-skinned beautiful young woman masseur and was disappointed to say the least when the masseur appeared. You may recall seeing the OO7 movie where Oddjob is a baddie. He was an enormous fellow, built like a tank with a penchant for skimming a bowler hat with a steel edge at James Bond. It was Oddjob's brother that now stood before me. I lay on the table and he began to pound me into a jelly. Each time he laid a hand on my back he emptied my lungs of air. My legs were not just pummelled, he tied them in a reef knot. I tried to call him to stop but I couldn't speak. Eventually, after what seemed like several hours, he stopped. I lay still trying to find out if I was alive and if my limbs functioned at all. I crawled off the table and slowly hobbled away. After getting my limbs to work more or less properly I got to the bar for something to help my recovery that I thought might take several weeks.

I had a totally different experience when I went for a haircut in one hotel. I'd noticed one fellow attended by no less than three young women – while one cut his hair the other two attended to his finger nails. After my haircut I was asked what else I wanted – massage? "Yes, please," I said. There followed the most delightful, almost orgasmic experience I've ever had with my clothes on – a facial massage. The young woman massaged my forehead, around the eyes, cheeks, scalp... it was truly wondrous. I was groaning. If you have never experienced this, then try it. Unfortunately I've never found anyone who could quite replicate that experience, but I'm still looking...

Singapore was another business area, and again we used an agent there for a time. I once travelled from Singapore into the Genting Highlands of Malaya to run a course. I shared a hired car with a well-dressed woman who was going to the same place – a gated resort on the top of a mountain. She was going there to visit the casino. We passed through what had been beautiful forests, now stripped bare by the chainsaw. It grew dark, and suddenly the car stopped. "Roadblock," said the driver. A gun came through the window and demanded our papers. It was the military, checking for bandits and terrorists. A foreigner had been murdered on the road a few weeks previously, I was told. At the resort I was shown to my room. It smelled strongly of damp – on inspection I found that the carpet was full of mildew. The resort was an oasis in the jungle but when I enquired about going for a walk I was told that it was forbidden – there was too much danger, they said. I rather thought that they were exaggerating until I read that a local had been eaten by a crocodile while washing in a river a few weeks earlier. Maybe I would just stay indoors! The casino was very popular with the wealthy of Singapore who would take a helicopter to get there. Unfortunately the helicopter crashed into the jungle a little later killing all on board and that was the end of the flights. The resort has been much extended since and a cable car now runs to the top of the mountain.

I ran a course in Milan for one group, who, I was told beforehand, could all speak English. In the event only half of them could, so that course went a lot more slowly, as everything I said was interpreted for the benefit of the non-English speakers! I had taken Lucy with me, as she had some free time and she was of the age to explore by herself. She experienced the dark side of Milan when her bag was slit open as she sat on a crowded bus and her wallet was stolen. Despite showing great initiative by jumping off the bus and stopping a police car the culprits got away. But we had a good week there and the company sponsoring the course gave a party for us at the end, and the delegates made a great fuss over Lucy.

Another European city where I ran a couple of courses was Munich. Hewlett Packard had an advanced technical development centre there and they had some very bright people. Some of the delegates showed me the city one night and we visited a couple of beer cellars and had a good evening. As I was running a course there the following week I had a weekend to look around so I rented a car and took off. Just ten or so kilometres from the city I found what looked like a pleasant country inn to stop for lunch. However, the atmosphere was cold and foreboding, the few people there were most unfriendly. Stag and boar heads lined the walls. I left quickly with an odd feeling. A short distance down the road I came to a crossroads where I stopped the car in the middle of the road. I had seen a signpost pointing to 'Dachau'. Although Dachau was a medieval town the name to me was only known as the death camp, the longest serving concentration camp of Hitler's Germany where more than thirty-thousand prisoners were murdered and hundreds of thousands were held and worked to death. My blood froze, I couldn't move for some time. I decided that I couldn't visit it although I thought that I should. This quite spoiled the weekend as I couldn't get it out of my head that the horrors had been happening just down the road from Munich.

The life I was now leading was hectic. I was travelling either within the UK or very often flying overseas, always packing and unpacking a suitcase, getting home and planning immediately to go away the following week. Staying in hotels rapidly lost any charm it may have had, and I became all too-used to eating alone in restaurants. I was not at home enough, and I was missing being with Jennifer and Lucy, who was growing up fast. While away I had to be self-reliant and deal with any situation that arose. I would always be faced with sixteen to twenty people and everything had to work smoothly. One time in Saudi Arabia in a major hotel I checked out the seminar room that I was to use the next day. The power point didn't seem to work for my projector so I requested the hotel technician. He was not available until the morning but I was assured that he would be able to fix the problem. Having experienced similar situations I knew better than to put my trust in the hotel, so I took a screwdriver from my case (I always carried one or two such essentials) and removed the power point cover. There were no wires attached or indeed to be seen! I managed to find an extension cable to link to a working power point. Another time in Hong Kong my suitcase failed to arrive with me at the airport – this happened at least half-a-dozen times on my travels. This time I went to the street market to buy clothes and essentials to tide me over until the case arrived. A tailor measured me up for a suit and had it delivered to my room the next morning! The case arrived several days later, having been around the world. Qantas paid for everything. I always carried the essentials for the training course in my hand luggage for just such an event. When my suitcase – one of those indestructible things that you can use to toboggan down a mountain, according to the adverts – came off the baggage conveyor with a hole in one corner and a crack across one side, I didn't even notice until I had the thing in the hotel. Only then did I see that I could put my fist through the hole! Yet again, Qantas coughed up. So take a bow, Qantas, although you do lose and damage things, you don't argue about it.

I had always had notions that our company should be able to sell consulting services – and we tried. We never really cracked this one, although I managed to get a few projects. They sometimes fell out of a project management course – if one of the attendees was particularly impressed with our apparent worldliness on all things pertaining to successful computer development projects, and fed back this information to his or her boss, then we might be invited to call in for a chat about a problem. One day in the office I took a phone call from a managing director who wanted such a chat. I agreed to drive to his factory in Slough for an appointment. I had two concerns with this. We had never had any people from this company – a pharmaceutical business that had grown mainly on the success of their famous eye drops – on any of our courses, and the situation the director described was remarkably like one of our case studies that we used on our best-selling course! The M.D.'s initials were the same as those of our case study as well! I thought that I was being set up, but I did some checks on the business and it all seemed to be kosher. I met the man, a charming chap and I agreed to do some work for him, which entailed an investigation and making a proposal as to whether they should 'go it alone' with a new computer and systems, or to stay with their parent company's systems. They were receiving a very poor service from their parent company and I went to the computer centre to see what could be done. After this visit I received a phone call from a senior manager there who threatened me that if I recommended anything other than staying with the status quo I would be in trouble – they would withdraw their patronage of our company! I was amazed to hear this and asked him did he really think that I would recommend anything other than what I thought was best for my client? I took no notice of this idiot and after a detailed investigation I proposed that my client should purchase a computer and an off-the-shelf system that would meet their needs. I had looked at various computers and software that would suit and I made specific proposals that the client accepted.

Another time I was asked to look at a systems development department that didn't seem to be working well, according to the new director who had recently taken responsibility for it. It was a well-known international engineering company, and I duly went along to take a look. I interviewed the data processing manager who was a classical nerd. Within two minutes of meeting him he was trying to show me some incomprehensible computer code! He was very proud of this, but that was not his job! His job was to produce results for his company, and outside of his glass office I could see thirty or more very disgruntled staff. I interviewed the senior people who all gave me the same story of terrible management. They never knew what their priorities were, the manager never talked to them, he just left them notes on their desks. When I cautiously raised this with the manager, he said that he used the 'management by surprise' system. I said that I'd never heard of this. "Yes," he said. "I leave them a note telling them that as of today they are on a different project, or that they should change priorities. Then next week I change it again. It keeps them on their toes." I was clearly talking to a seriously disturbed person, or one who had no idea about management. It only took a day here for me to be able to tell my client that he had no option but to get rid of the manager and to hire one who knew what they were doing. I gave a brief verbal report and followed it up in writing, but it was a crystal-clear case. I would have loved to get in there myself to sort it out. I knew just what needed to be done – I'd done it at Plessey! This sort of situation was not that unusual then – senior managers and directors were often totally out of touch with the computer scene and thought that they had to hire a technical expert to manage things. Of course the I.T. manager has to understand the technical issues but overall they have to be managers of people and situations and have to understand how to provide solutions to their clients' business problems; their detailed technical knowledge is secondary. This was very poorly understood at the time. The problem was compounded by technical people who were moved into management but who then insisted on staying at the front of the technology and continued to work at the detail level, ignoring the managerial issues. The technical issues are just too much fun to delegate, for these people!

One of the first consulting projects that I became involved with was a timber importing company. It was a small, private business and somehow we were invited to look at it. As was common then, the business used a small computer for the basic accounting processes. The real issues that the business faced was dealing with the complications of importing, storing and selling the timber, with its masses of paperwork. The directors realised that they were drowning in paperwork that it was becoming increasingly expensive, and perhaps more importantly, that management of the business was nigh-impossible due to the lack of meaningful management information.

Two of my colleagues performed the initial analysis of the systems using the systems analysis tools that we taught on our courses. The processes were labyrinth-like and the data did not seem to lend itself easily to analysis techniques. In fact someone from one of the 'big-four' consulting companies had recently written an article saying how it was impossible to use a computer system in the timber industry as the variations with the timber made any sort of codification infeasible. It was indeed a complicated business, involving overseas timber agents, ships and shipping contracts, keeping track of timber in transit, in docks through to warehouses and timber yards. Then there was the selling function – almost as complicated! But my colleagues did a brilliant job and the processes unfolded and the data sets became visible under the close examination of the analysis techniques. I became involved when the analysis was reaching its conclusion and the next step in the project loomed. I had to quickly get up to speed with the work before taking on this next step – to convince the directors that we had got to the bottom of their issues – and then to progress to a solution. I made a presentation to the the board – just three of them – when I ran them through our findings. They were amazed and delighted that we had documented all of the processes and agreed with our findings when we nailed the basic issues. They had never seen the whole business laid out in the diagrammatic way that we used and they understood, for the first time they said, why they had particular problems in the business.

It took little time for them to agree to our proposal for the next stage – to go out to tender for a software development company to develop the system, and another tender for the hardware that would be required. We were contracted to deal with this and I went through the process, eventually making a recommendation to the company which they accepted. The winning software people produced a fixed-price contract (almost unheard-of in the industry due to the unknowns), saying that our request for tender was the best-documented and clearest that they had ever seen. Again the client asked us to lead the way through the development and I set up a committee which I chaired when we held regular meetings with the software development team leader and the operational managers from the timber business. Setting an implementation date became an issue – the company M.D. set a date which I questioned as I could see no logic in it. During a rigorous discussion he eventually sheepishly said that the date was his birthday! We agreed that there were more important criteria. Pressure then was put on to have the system in by Christmas – a very busy time of year for them – not because of the usual reasons but this was the time of the coming of the Russians! It was all due to the ice-breaking ships clearing a passage through the ice and delivering the large stocks of Russian timber. I saw that that this date was highly risky and stressed the need for implementing the system in a quiet business time in order to minimise business failure. This argument won the day. I also planned to implement the system in phases as it was going to be a steep learning curve for the staff and they would need good support during this difficult time. I kept the directors informed the whole way, reporting progress against the schedule and budget. It was all very exciting, as the system, if successful, promised to revolutionise the way that the company bought and sold its timber.

The system was developed over a period of almost a year, the system appeared, the hardware was installed and once the dust settled it worked exactly as it was designed. The board of directors were very happy with the result, and the business prospered. In fact the timber industry went through a difficult time some years later but this company was able to adapt and sailed through, and exists to this day. It had been a most valuable exercise for us, giving us great experience with the tools and techniques, so much so that we adapted it and used it as a case study on our courses for many years.

During this time Jennifer and I were feeling wealthy and we had bought a flat in Marbella, not far from Gibraltar and we went there a number of times, firstly to furnish it then subsequently for breaks. It made a good holiday home but eventually was to prove to be a problem. We had good sea views at first but then the developments began to crowd in. The flat was brand new and on our first visit I found that the power points weren't working. The electrician arrived and proceeded to pull out all of the wires from the power points! I tried to indicate that I thought there was a problem with the fuse box. He ignored me and continued his task until he sat on the floor surrounded by wires. He scratched his head and only then looked at the fuse box. There was no connector between the input wires and the internal circuit breakers! The next thing was that the occupants from the flat below complained that water was running down their bathroom walls. Nothing appeared to be wrong that I could see so again we called the agents. A day or so later men were digging up the drains outside. It transpired that there was no connection to the sewers from the entire block and that the drain water was just backing up inside the building! My parents and my sisters went there and enjoyed it but there was often a problem with the plumbing or the electricity. The bank there was hopeless and often didn't pay the standing orders for the utility bills so the power or gas would be cut off. When I confronted the bank (phone calls got me nowhere) the bank official just listened to me as I told him that they had messed up yet again and he agreed with me that they were indeed hopeless!

I enjoyed visiting Northern Ireland and Eire. In Belfast I gave a course at Stormont Castle, which had been the seat of government and was now used by MI5 among others. As it was in the time of 'The Troubles' there were soldiers in flak jackets looking out from their sandbagged sentry boxes, with guns at the ready. I was a bit nervous walking from my hotel to the Castle as the road had been the scene of a recent shooting. How I ever agreed to do that job I don't know. It was much more relaxed in Dublin where I gave what must have been a couple of my best courses. We were in a rundown hotel almost on the beach in Dublin Bay that reminded me of *Fawlty Towers*. The fabric was crumbling but the staff were lovely and the food excellent. Apart from the vegetables, that is, which were boiled to buggery. This is common there, I was told. The delegates enjoyed the course and worked hard. Somehow everything went exceptionally well and we all had great fun. It was a course for Price Waterhouse who were one of our best customers. We had designed special courses for them in London and subsequently presented them in many countries. I was impressed with their staff – they must have had an excellent recruitment programme because they only took on very bright people. We had some riotous evenings and they took me to a number of pubs where the fiddlers played, people sang... and I got sozzled. I don't remember getting back to the hotel. I ran another course at the Guinness brewery in Dublin and this was a corker as well. I was amazed that every worker was entitled to a bottle of Guinness with their lunch! I would have been useless for anything in the afternoon had I taken advantage of this but I noticed that many of the staff did so. I requested and obtained a visit around the brewery and saw the huge stainless steel vats where the black thick liquid was brewing. "We had a sad accident here recently," said my guide. "The night watchman was found drowned in a vat of Guinness. At the autopsy," he continued with a straight face, "they established that he had got out three times for a pee!"

One day I was in the Welwyn office when I took a phone call from Phillips Petroleum, one of our regular customers. "We'd like course X," said the chap, "around this date." "Can you manage that?" I looked at the schedule. "Yes," I said, "we can offer you a couple of dates around that time." We agreed the dates. "Who will present it?" asked the chap.

It was only then that I asked if the location was to be their usual place – a computer building in a grimy south London suburb. We had a rule that whoever took business on the phone had the first refusal, so location was a consideration. "No," came the response "Stavanger."

"Stavanger, as in Norway?"

"Yes," came the reply.

"Then I'm doing it," I said.

I arrived at Stavanger and booked into my hotel. I had a large room with a lobby where one could hang wet clothes and boots. I had showered and was only wearing my jocks when there was a fumbling at the door. I heard the outer door open then the inner, and a tall, well dressed good-looking blonde woman appeared. She and I looked at each other. "Was this part of the deal?" I was wondering, while being aware of my state of undress. Perhaps it was part of a "welcome" package? Or was this some Norwegian custom that I hadn't been told about? Just stop your mind racing ahead, this is not fiction! Eventually I asked her feebly, "Yes, can I help you?" She spoke perfect English and said that she thought that I was in her room. It transpired that she was on the floor below mine, and she had read the post code on the key ring ('If found return to…') which was my room number! And her key fitted my door! Umm! When she had left I thought that I should have invited her for dinner, but in the dining room later I saw her with a rather large Norwegian man so I merely smiled as I passed her table. I had Saturday free before meeting my clients so I took a look around the small city, wandering the narrow streets. It was quite charming. I was looking in a shop window when I became aware that the city noise had diminished. Looking up I saw that I was the only person on the street. "Closed" signs had appeared in all the shop doorways. I was perturbed. What was going on? Was it a practice nuclear air raid? Where was everybody? It had all happened so quickly. It was midday, and Stavanger had closed down for the weekend.

The venue proved to be a lovely hotel situated, as they say, on the beautiful Stavanger fiord. I met the Phillips delegates and we travelled from Stavanger by chartered coach to the hotel, making a ferry crossing en route. Crates of beer were loaded onto the coach and shortly after we'd started some fellow was bringing the bottles around. I took one and within a few minutes was offered another. I shook my head. "It's free," said my benefactor, "take it." I declined to his amazement. All of the others were drinking as fast as they could. When we arrived at the hotel they poured off the coach. Many of them didn't make the evening meal. I learned that alcohol was difficult to buy in Stavanger, having only two outlets and one of those looked like a Co-op funeral parlour. Drinking in the street was unlawful and this was strictly enforced. So once these chaps were free of the restrictions... I had to lay down a few rules on the first day of the course (after I had rounded them all up) and we had no problems after that. The food was marvellous, and included a great smorgasbord. On the last evening we had a bit of a party. Some of the chaps decided to take a swim in the fiord, stripping off and jumping in from the hotel jetty. Glaciers were entering the fiord just a short distance away and a crack could often be heard as another chunk of ice fell in. They weren't in the water for long.

Jennifer came out to join me at the end of the course. We hired a car and drove to the hotel that I'd been staying in and explored the surrounds of the beautiful fiord. Then we drove further north. Some of the roads had just been opened and the snow was piled up to twice the height of the car on either side of the road. Waterfalls tumbled down the mountains and sometimes crashed down alongside the road. It was enchanting and made a very pleasant break for us both.

For some years Jennifer and I had been involved with running a club for adults with learning difficulties, (the term used then was 'mentally handicapped'). It was the Dowsett Club, which met at a church hall in Palmers Green in north London every Thursday evening. It was managed by the Enfield Society of Mentally Handicapped Children, part of the National Society, whose president was Brian Rix, the actor. We were regular 'helpers' for some years, running activities for some thirty or so people of all ages, mostly thirty-years old upwards, one or two in their sixties. They mostly had mental ages of eight to fifteen-year olds, and some had multiple disabilities, some had Down's Syndrome. The woman, Nora, who led the club was a wonder, and with her husband did a fantastic job. Jennifer and I became quite involved, and the four of us became good friends. A few of the members were married couples, and one couple had children. Another woman (I'll call her Maureen) had a baby. The difficulties encountered in day-to-day living were numerous. Although the children were loved, it was not enough. One winter evening I left the clubroom and went to my car. Maureen sat her tiny baby on the ice-covered roof of my car while she put on her own coat. The baby, unsupported, was about to topple off and I was just in time to catch it. Just an example of the everyday things that our members couldn't really manage. It was made more difficult when government policy changed and the institutions that many of the members were in were closed down. The inmates were turfed out and would "take their place in normal society". Unfortunately, society wasn't ready, willing or able to accommodate them and they fell into the cracks. I visited a couple with two children and was horrified at the squalor. A baby lying in its filth on a filthy bed, a flat full of dirt and rubbish, no food... Another couple had an Alsatian dog in a tiny flat. They kept it in a room, rarely taking it out. I walked in and was almost sick with the smell of the dog mess covering the whole floor. The bedroom was a foot deep in dirty clothes – they couldn't wash and dry them so they just wore them then threw them down. More clothing was provided by the social services... One woman had a broken arm that had not been attended to and I established that her husband had pushed her down the stairs. She had been found wandering naked through Palmers Green shopping centre. We did what we could in these situations and were called upon by the police from time to time to assist. The social services were totally inadequate for the supporting role required and the support that organisations like ours could offer was limited. Some of our members were always getting into scrapes with the law and the police did not have the training to deal with them.

Apart from the social aspects, part of the club's *raison d'être* was to give the members' parents or carers a break. We organised day trips and I obtained a licence to drive a minibus to take them around. We obtained a Second World War ambulance which was converted to be a 'people mover', this was a huge vehicle and fun to drive. It was a Daimler, with eight gears and a long gear lever to go with them. Driving through London traffic I found that cars quickly moved out of my way - whether this was the aggressive look on my face as I wrestled with the gear box or the large steel bullbar on the front I don't know! Each year the club members would be taken on a week's holiday, alternating between the UK and somewhere overseas. Jennifer and I took part in some of these, taking them to Skegness and Blackpool and to Lisbon. The members of course had a wonderful time. Lucy came along on these trips and she was very popular with the club members. She developed a good rapport with them and although the whole scene must have been a little disturbing at first, she quickly adapted and became very helpful. Some of the members had physical issues and were overweight or obese. One such woman became stuck in a bath on the Lisbon holiday and Jennifer and Nora were trying to get her out. She was firmly wedged, with the water drained out in front of her and backed up behind her. Shrieks of laughter were heard coming from the bathroom as they all struggled to budge her. Listening to this outside I called for help and more hotel maids appeared who also quickly succumbed to fits of laughter. She was eventually heaved out with loud suction noises and water everywhere. It was the talk of the hotel for the rest of the trip. Situations like this were not uncommon and provided much laughter and many dining-out tales. I took on the role of club co-ordinator for a time and liaised with all the clubs in the Enfield area. I then took over the role of managing the club from Nora, who wanted a break. I was also asked to take on the chair of the Enfield society but I turned this down as by then other events were taking priority with me.

Years later, when in Australia I was being interviewed for a job the 'career consultant' advised me to leave this mention of voluntary work out of my cv. He thought that it had negative connotations! What a berk, I thought, but it showed that there remained much ignorance and prejudice about the mentally disabled.

CHAPTER 8
A Difficult Time

At home in Cockfosters my relationship with Jennifer was suffering. This was hardly surprising as I was away so much. Jennifer had long harboured frustration over her career – or the lack of opportunity to pursue it as I had mine. She had returned to work once Lucy was of an age where this was possible, but she felt that she had been held back, which was true. Our relationship deteriorated. On Lucy's ninth birthday we were on holiday in Wales when I experienced a sudden noise in my right ear – the noise that one hears on putting a sea shell to one's ear – tinnitus. The local doctor told me to go to a specialist on my return home, but he thought that it was Meniere's Disease. He proved to be correct, and there was no cure. At first I suffered from nausea and dizziness, for a time I couldn't walk a straight line but these symptoms slowly disappeared leaving me with a deaf ear and the noise. I have lived with this ever since, and I have often wondered whether the strains of our relationship contributed, as the affliction may be caused by stress. The nerves connecting the hearing mechanism to the brain are damaged, so hearing aids, I was told, were of no use. However, I have learned to more or less cut out the noise mentally, otherwise it would have driven me quite mad. Much later, at nearly seventy, with poor hearing in both ears, doubt was expressed about the diagnosis of Meniere's Disease, but no matter, nothing could be done anyway. Except that I did buy two hearing aids – with some beneficial results.

So it was that on one course that I ran at a hotel in Hertfordshire I met this young woman, Gina, who really attracted me. I could feel the sparks fly. I thought that nothing more would come of it, but it did set me off and made me realise that I was in fact very unhappy with my home life and that I really needed to do something. Lucy was in the sixth form by now and I was aware that she was hurting. I was losing out with our relationship at an important time in her life.

And that was the beginning of a long and painful episode in my life, when I struggled over my future. It got so that I could think of nothing else but what should I do? I went to a counsellor who wasn't much help, but then I found another who was a bit better. I was clearly not myself when I turned up to run a course one Monday morning at a country mansion not far from home. A colleague of mine was at the door. "What are you doing here?" she asked. "You're at the wrong venue!" I should have been at another hotel, some miles away. Phone calls were made to excuse my lateness as I drove to the correct venue at The Thatched Barn at Boreham Wood. There were twelve delegates sitting there quietly as I walked in. I made some apology and began the course. By now I could have given it in my sleep, which was just as well for I was on auto-pilot for the whole week. It wouldn't have been one of my better courses. As soon as I stopped speaking I would look out of the window and thoughts would crowd in on me and I was unable to think of anything else. I felt awful about the course as well. I had never before been late, in fact I always made a point of having an hour or two to spare to ensure that everything was set up correctly, if I wasn't there the night before. But I got through it and the delegates' assessments weren't disastrous.

This time was tough on the four of us. Our Cockfosters house was sold, we bought a flat in Swiss Cottage and Jennifer and Lucy moved in. I moved in with Gina. This only lasted for a short time before I found a place of my own, but in practice I spent much of the time at Gina's, when I wasn't away working. The workload had not eased and I travelled around from week to week, country to country. I wanted a divorce, but that wasn't straightforward. Gina was becoming impatient. I was stressed out and succumbed to glandular fever. I was laid up in bed with pools of water on my chest each night. I changed flats several times so I was living out of a suitcase even more than ever. I put some belongings into store. For a time first one, then two of my colleagues needed a place to live so the company rented a house that was used partly as an office, and partly as accommodation for three homeless presenters! This eased my pain a bit for some months.

Gina's patience was wearing thin. She decided to move away to leave me to sort out my life. "See me when it's all sorted," she said as she went as far away as possible – to Australia! Jennifer also decided to move and left the flat in Hampstead for a new job in Coventry. Lucy, after a difficult time in the sixth form, had left for Edinburgh University to study fine arts. (She had switched from the science stream to the arts stream, which involved doing 'A' level art in one year – which she did successfully.) She did a five-year course at Edinburgh and obtained an M.A (Hons) degree. I lived in the Hampstead flat for a time but then I discovered that large cracks had appeared in most of the walls – the house was cracking up. We owned the basement flat (with a garden) in a large Victorian semi-detached house. It had to be underpinned. This was a situation that I could have done without but I had to manage. The body corporate, consisting of the other flat owners, were not interested in the underpinning. Their flats were not showing the symptoms, (but would eventually) so I had real trouble to get them involved. Architects, surveyors, structural engineers, assessors, insurers, builders, the body corporate, and other specialists were involved. At one time I was communicating to twenty different bodies. It was a nightmare. I had to move out and large trenches were dug inside and outside the building and reinforcing and concrete was poured in. Most of the flat needed to be repaired, plastered and redecorated. It took over a year and cost more than £40,000. I still have the files – the paperwork filled two large ring binders. This was still going on when I went to Australia to see Gina. It was becoming clear that she had her doubts about my long-term intentions and she was beginning to meet other men. I hadn't told her that I was on my way and so arrived unannounced on her doorstep in Melbourne. She was surprised to say the least! She looked through the peephole and wondered for a moment whether to open the door! We are both glad now that she did! I stayed there for some months while I researched the possibilities for permanent work there. I paid the London architect to oversee the work on the flat in my absence. He was the senior man at his practice and seemed to be a very experienced fellow. From the lack of news from him and the lack of responses to my questions I feared that something was wrong. The internet had yet to be available so communication was by fax and letter mainly, as the time lag made telephone calls in business hours all but impossible. On returning to London I inspected the flat and found it to be a disaster area. Nothing was being done, a wall had been demolished and the flat was partly open to the elements – and to vandals, and what work had been done was not to the specification. I rang him to make an appointment and I was told that he wasn't available. Muffled noises in the background aroused my suspicions and I immediately drove to his company office in St John's Wood and went in. His secretary said that he wasn't in but I burst into his office and there he was, cowering behind his desk. By now I was furious and I must have looked threatening (I find this hard to believe myself) but he showed me his arm in a sling and whimpered "Don't hit me, don't hit me, Mr Brown, I have a broken arm!" He was pathetic. I felt like throttling him. "I trusted you to look after my interests," I shouted. "You're a disgrace!" He had little to say for himself by way of explanation; it seemed to me that he was either totally incompetent or he was having a breakdown of some sort. I eventually made a complaint to the architects' body that investigates complaints of this nature and it transpired that he was its chairman! How's that for a bad choice of architect! It was an involved case, but they did put someone onto it and I got an apology and a little compensation. I hope that he was dismissed from all duties, but I don't know that he was. It took more months of slow progress to complete the work at the flat. I had to manage it and chase it every inch of the way and eventually it had a satisfactory outcome. When at last I moved back in for a short time the final straw was that a water leak developed in the sitting room and the concrete floor had to be dug up – again – for the pipe to be repaired!

Another incident at the flat occurred when I was living there and two flats were put up for sale. Several agents were involved, as is common in the UK, so five large agents' boards appeared on the gatepost, securely fixed with wire. This looked ridiculous and was quite an eyesore. After a little time I rang the agents and asked them to remove the boards. "We need the flat owner's permission to take them down," one said. "No you don't," I said "You need my permission to put them up. And you haven't got it, so take your board down." After a short discussion it was obvious we were getting nowhere. "You have until 10 a.m. on Saturday to remove your board," I said, "otherwise I shall remove it and chop it up." (By now my patience was thin!). I repeated this ultimatum to the other agents who were equally unresponsive. Of course, nothing had occurred come Saturday morning so out I went with my tools. The wire they used to secure the posts was extremely thick fencing wire that ordinary pliers wouldn't touch. But I had a pair of ex-USA army wire cutters! Down came the boards, and with my electric saws I cut up the boards and the posts, depositing the pieces in the various tenants' dustbins. Several neighbours gave nods of approval as they passed. I never heard a squeak from anyone about this.

One of my protagonists in the flat saga was Camden Council, who claimed the Thatcher poll tax from me. As the flat had been uninhabitable for more than a year and hadn't been lived in since the tax had been introduced, the poll tax was not legally due and I explained this in writing to the council. They merely repeated their demand. This went on and despite several phone calls the demands kept coming. In desperation a few days before I was due to leave London for Australia I called at Camden's offices. The offices were closed as the staff were on strike! I kept receiving the demands and then the matter was passed to a debt collector. I received threatening letters and again my reasoned responses were ignored. It really was most frustrating dealing with entities that did not seem to take any account of my explanations. Not once did I ever get the notion that my letters had been read. The final straw was when the debt collectors said that they were going to take possession of my assets! By then Jennifer had sold the flat, I was in Australia so I said "Good luck" to them. I threw away the letters from them after that without opening them.

The Hampstead flat was a long saga that was going on in the background while I tried to get my life in order. Jennifer had agreed to a divorce and that was proceeding, slowly. The day came when we both went to Warwick for the court hearing. I had to travel from Melbourne for it. Her lawyer was fighting for as much money as she could get. Naively, I had thought that we would just split our assets 50/50. Not so! My lawyer turned out to be useless, (another poor choice!) he was outwitted at every turn. The end result was that Jennifer got the Hampstead flat, worth quite a sum now that it was finished, and most of our money. I got the flat in Spain, worth little and with a large loan on it! At least this was one chapter coming to an end.

There was one amusing incident in the middle of all this. At one time I had rented a small house near Hatfield and I wanted to rent a television. I was planning to leave for Australia in three months, so it was a short-term rental that I wanted. I explained this to the young rental shop assistant. She began "The six-month rental is... but if you keep it for twelve months...". "No" I said, "I only want it for three months". She began again "The six-month rental is...but if you keep it for twelve months..." I looked around the empty shop. "Are we on Candid Camera?" I asked. She looked blankly at me. "Look" I said, "I'm going abroad in three months. THREE MONTHS. After that I shall not be here. I shall be overseas. I just want to rent a TV for that time. Can you do that?" "The six-month rental is..." I was sure now that this was some kind of set up and I looked around for the camera. I was reminded of the dead parrot sketch of Monty Python. I gave up eventually and found what I wanted elsewhere.

During the early days of our relationship Gina and I took a skiing holiday in France. Gina was quite an accomplished skier but I had never put on skis, ever. So I had a week of ski school and learnt the basics, becoming quite bruised in that time as the snow was pure ice for the whole week and every fall – and there were many – was like falling onto concrete. But I had great fun. The second week we skied around together, Gina leading the way saying "Just follow my tracks". I did the best I could. At some point I noticed the signs on the tracks we took all had black diamonds on them. "Aren't these the most difficult runs?" I asked. "Yes" she said, "it's good practice for you!" We did every black run in the resort. On one run we reached a long, very steep drop with a 'pipe' bottom and an equally long steep rise the other side which looked to me to be almost vertical. "Lean forward" Gina said, "And you'll be alright!" I hesitated for a moment, but knowing that the longer I hesitated the worse it would appear I pointed my skis and went! To say it was exhilarating is an understatement. I was so relieved to arrive at the bottom in one piece that I then forgot my instructions. I went up the reverse slope and became horizontal, with the inevitable result. I arrived back at the bottom with skis, poles, hat, goggles scattered in profusion. I became quite the expert at getting back onto my feet.

On one of my visits to Australia in 1989 Gina went away on a pre-arranged trip. One of my scheduled courses was cancelled at short notice so I had some time up my sleeve. I booked a trip to the outback, from Alice Springs up to the Top End. This was essentially a camping bus trip, not really my thing, but I didn't have a lot of choice. We traversed the Tanami track, getting bogged in bulldust, and just about wrecked the Mercedes bus on the rough roads requiring emergency repairs to be carried out in Darwin before we could move on through the Kimberley. At one bush campsite we were sitting around a roaring campfire under the beautiful star-filled night sky when I was reminded of my boy Scout days when we would often sing around the camp fire. I must have had a glass or two of wine as I suggested that we sing a song. The group agreed readily so I led them in "Camp fire's burning". It lends itself to rounds, so we sang it first in two rounds then in four. They all loved it. I often recall the faces in the firelight, the sparks drifting off up to the stars as we enjoyed the moment. From the Kimberley we began to head back to Alice Springs. By this time I had had enough of traveling by bus – although we were covering great distances we weren't doing sufficient walking or exploring. So I planned to leave the group, hire a minibus and do my own thing. A woman asked if she could join me, and several others had the same idea. On my last night with the group we went to a pub in Alice Springs, had a meal and a bit of a party. One chap wrote a poem in which he mentioned each person in the group, and extolled their virtues or some special characteristic. It was very well done and he was very funny. A hot-air balloon trip was planned for early the next morning and immediately after this I planned to leave, collect my vehicle and head off to Kings Canyon. As the evening was drawing to a close I checked my plans and realised that I was going to be short of time if I did the balloon trip, so I went to the notice board with its list of thirteen names and crossed mine off. "Oh, good!" said the chap next to me as he moved his name up from the waiting list to take my place. I said my goodbyes to the group as the balloon party were to be up at 5 a.m. Next morning I was packed and just about to go into town when a mini-bus roared into camp and dropped off one of our bus party, a young woman. She was hysterical. It seemed that two balloons had taken off together, and one of them had risen directly under the basket of the other. The balloon was ripped, and the basket with its passengers dropped like a stone more than 2,000 feet to the desert. She thought that it was our party but she wasn't quite sure. I stayed at camp with a largely silent group while we waited for more news. Eventually the police arrived and confirmed that it had been our party. All twelve of them died, and the pilot. They had had just more than ten seconds sitting in the basket, knowing that they were going to die. They were found sitting around the basket with their arms linked. I had got to know some of them quite well in the ten days or so that we had been traveling together. They included a young couple recently married, the poet of the party (who had taken my place), and a man who had been unable to persuade his wife to join him in the basket. She was very calm and quiet as she waited for the news, while the rest of us tried to offer support.

The young woman who had been hysterical had actually climbed into the basket, but the pilot checked the number and asked for a volunteer to get out as only twelve were allowed, not the thirteen we'd been planning for. I left the group much later than planned that day, and it made the remainder of the trip very sobering. I kept thinking of that evening singing around the campfire. I couldn't help the thought that I was so very nearly in the group. I decided then to ensure that I lived life to the full as one never knew what was around the corner. No more lingering over decisions, I thought, just get on with it! It was several years later that an enquiry into the accident blamed one of the pilots and suggested changing procedures when two balloons took off close together. It was not until my seventy-fifth birthday that I eventually got around to making my first hot-air balloon trip. I was somewhat nervous as several balloons took off together. I explained my concern to the pilot at the end of the trip. He understood my worries and told me that the pilot of one of the other balloons was the brother of the pilot who had died at Alice Springs. It seems that hot-air ballooning is a close fraternity.

Back in Melbourne another problem arose on top of all of this. I had been in Australia for some months initially and I had agreed with my partners that I would service all of our commitments there. Our agent was providing me with a regular stream of jobs. As well as this I began to obtain business on my own account, having set up a subsidiary of our UK training company. Over the years 1987-89 I was extremely busy running courses, mainly in Australia but also in Dubai and the UK. I ran some seventy courses with over seven hundred delegates covering some 250 training days in that period. By now we had seven partners in the UK business. One of them got in touch with me to let me know that the partnership was embroiled in turmoil, with the two original (now senior) partners trying to take over the total control of the business. It seemed that I had unknowingly been the bridge or communication between these two and the remaining partners, and that communication had fallen apart in my absence. Certainly when I was based in the UK we had all been getting on fine, and the business was running nicely. We all had capital tied up in the company, we were making good livings from it and it was in all our interests to maintain it. I returned to the UK to see what was going on. I got back into running courses there again while we tried to sort things out. The business was going very well, with a good revenue stream and good profits, it seemed that it would be madness to let it break up. The arguments went on for some months and the junior partners even reached the point of taking legal advice and having a meeting with a lawyer where myself (representing the junior partners) and Colin outlined the main arguments. Eventually a company appeared out of the blue and offered to buy us! It was a complete surprise, but most timely. We sold the company for £1.7 million, far more than any of us expected. We received shares in the purchasing company, and cash. I pocketed my cash and negotiated a job with them in Melbourne as training manager in their subsidiary there. So at last things were settling and my luck seemed to have changed! I returned to an impatient Gina and Melbourne, to a new job, with a car, and Gina and I found a house to buy together. Things were looking up at last! There was however, yet another cloud around the corner.

CHAPTER 9
And Now For Something Different

This had been a most disturbing period of my life. I had lived at seven different addresses in and around London and I had travelled back and forth to Australia seven times in a few short years. I was very much looking forward to a more settled life, but there were one or two hurdles to cross first. I settled in with Gina, living in her flat close to the beach in Elwood, Melbourne, while we found a suitable house. A cheque arrived in the post for a large sum, being from the sale of the business. It was for more than one hundred thousand pounds, I had never handled a cheque for such an amount. I immediately deposited it in a building society in a term deposit for three months, that being the time we thought we required to find our house. After much searching we found a very nice house in a green and pleasant Melbourne suburb – Kew – that was coming up for auction. We thought that the auction system was a complete con – at the time the auctioneer was legally able to accept bids from a passing dog, the lamp post – and so it was with this house. We didn't attend the auction, but we had a spy there, who reported that he thought there were no genuine bidders, and the house was passed in. There was a risk, of course, that the house would be sold at the auction but we gambled that it would not sell due to the poor economic climate. So I was able to negotiate with the agent after the auction and got the house at our price. We had one week until the term deposit matured (what timing!) and that would then pay off a substantial part of the house purchase price. I put in my letter to claim my money. The next morning the headline of the newspaper told me that my building society had collapsed, no money was forthcoming! It was the Pyramid Building Society, and was one of a string of financial disasters that would envelop Victoria and cause the resignation of the Victorian Premier. We were mortified – Gina had deposited a large sum as well, we were about to buy a house and now we had no money! We had deposited our money with little or no research, and the one relevant fact that we hadn't established was that building societies in Australia were not guaranteed by government, as they are in Britain. I went to the office in the city where a smashed window, a pile of unopened mail inside the door, and a policeman on duty told the tale. I approached the estate agent who had sold us the property and explained the position. "I have no money," I said, "I cannot fulfil the contract. I need to back out". He looked at me coldly. "You can't do that," he said. "You've signed the contract, you have to proceed. If you don't we'll sue you!" Mortgage interest was standing at eighteen per cent and we had to negotiate a huge loan.

But this wasn't the total of our pain. I was now working in my new job in Melbourne as the training manager for the company that had bought out the UK business and I had been with them for a few months. The economic climate was terrible, and the training business was suffering. A week after the Pyramid collapse I was called into the manager's office one Monday morning and told that I was being "let go", as they were shutting down the training division entirely. So now I had no job on top of our other worries! Australia was suffering from an economic depression and was not looking so good after all

But every cloud has a silver lining as the expression goes. Gina was earning good money as a contractor and we managed on her income while I tried to find work. I applied for more than fifty positions, got a few interviews – but no job. It was not a good time to be looking for employment and as I was fifty… but one interviewer rang me to suggest that I should meet a friend of his who might have something to interest me. I met the chap in question, and the upshot was that we started up a training company. (I wasn't able to restart the previous subsidiary company that I had been running as it was an element of the company sale). It was a terrible time to be starting a new business, but it seemed that I had little to lose. We started the company in a spare room at home, with me working full-time and my partner working part-time. I had to write the training courses to suit the changing times and the Australian environment, but our main worry was finding the customers! I made some visits to the large financial organisations in Melbourne and sold some courses to National Mutual and to WorkCover. I also obtained a lucrative consulting job from a client and that really helped. Slowly, slowly we began to win more business. Although I had been running courses in Melbourne under my previous business arrangements, our business name was unknown and the I.T. market was very tight indeed. I soon found that it was nigh-impossible to run courses and market and develop and do the administration so we took on a salesman at the earliest opportunity. Neither my business partner nor I took any income from the business for about eighteen months and it was a bit galling to be paying the salesman! He wasn't a great salesman, but we weren't paying him a lot! Although my intention had been to market only to Melbourne as I was fed up with travelling, it soon became apparent that we would have to cover both Sydney and Canberra. Eventually we also sold courses in Perth, Adelaide and Brisbane and elsewhere. Later still we ran courses in Hong Kong and Papua New Guinea. After six months or so we took an office in the city and took on an administration assistant as business was picking up. We moved again in six months to a bigger office and stayed there for some years. By now we were using computers and I had an Apple Mac at home. We installed P.C.s in the office and soon the internet appeared with e-mail and communications gradually changed. We were marketing with 'mail-drops' – sending advertising brochures to anyone who might be interested in buying our sort of training. The difficulty was two-fold – our target market was the software development staff of the larger companies, not all of the other IT technical areas, and the IT market wasn't as sophisticated and as well-understood as it was in the UK. Secondly, company organisation greatly differed from one to another, so that it was difficult to establish just who was responsible for the training of the software development staff. Training was a much-neglected area in fact, in Australian commerce and industry as a whole. It has improved but it still remains an afterthought in some organisations. We had to create mailing lists of the appropriate people, visit them, create a relationship... and hopefully sell some courses to them... before they left!

The federal departments in Canberra became good customers. For some years the Australian Tax Office was our best customer and I loved taking money from them! I always felt that every dollar was worth two! Australian Customs, Centrelink, Industrial Relations, Dept of Ageing, all these bought our courses and more importantly, booked in-house courses. So here I was again flying out on Sunday evening to stay in a hotel and run a course for four or five days.

The attitude of the delegates in Canberra showed a marked difference to those in private companies, especially in the Nineties. Working closely with the delegates on any course, I could quickly assess their working culture. The culture and enthusiasm of the delegates made a huge difference to the course, and Canberra was always more of a challenge to us. The Canberra conversation over the coffee urn would largely consist of discussion of their working conditions, pay, and promotion prospects. The actual work that they were being paid to do (and quite well paid I should add) was of minor interest to them. In a commercial company the discussion would be centred on the systems, problems, technical issues... and these topics would be discussed with energy. Two examples will illustrate the sort of thing that I mean. On our courses we always had practical work where we would split the delegates into teams and give them a realistic case study to work on. One day on a government department course we had reached lunchtime, and the teams were working independently in their private rooms. I had told them that they were to organise themselves and manage themselves as if this were the real thing, and that included when to break for lunch. I was the 'fly on the wall' in one room, observing how things were going. Three delegates were at the flipchart board, earnestly discussing the issue. The fourth delegate, a young woman, was sitting separately, reading a newspaper. I sat down quietly beside her. "What's happening?" I asked. "Why aren't you involved here?" She turned away from the newspaper reluctantly and looked at me, and then at her watch. "This is my lunch hour," she said coldly, and went back to the newspaper.

Around 2001 I was again in Canberra running an in-house course on site for a government department. I arrived as arranged at the building just before 8 a.m with the course materials – usually a large box containing the twelve or so binders containing the course manuals and other training materials. I found the correct floor with some difficulty – the signage was non-existent. The lobby contained a large quantity of old telephone directories, boxes and waste materials, stacked by the doors, which were locked. The usual start time was 9 a.m but I had to arrange the room and set up and I always allowed an hour for this. My contact had agreed that someone would be there to let me in. After twenty minutes someone arrived who knew nothing about the arrangement, but let me in. The next issue was to establish which of the several meeting rooms we would be using. Again it took some time to establish this. Eventually I was organised and ready to run. By 9 15 a.m I had about seventy per cent of the delegates there and I made a start, the others drifting in slowly. One didn't turn up at all. I was used to this sort of thing in Canberra and had made all the usual noises to our client about these domestic arrangements, which often fell on deaf ears. I had to work hard to generate some enthusiasm and to get the delegates onto my side and the course eventually finished well.

Before I left I collected all the excess paperwork and put it into my cardboard box, and left it by the wastepaper basket. I was back at the same site four weeks later, to repeat the course for a second group. The waste materials were still in the lobby outside the front door, and my box of waste materials from the previous course was where I'd left it! I had never seen this sort of poor housekeeping in the private sector. It seems that in the public sector the roles, procedures and responsibilities were laid down in great detail, and endlessly discussed, and if clearing the rubbish wasn't part of your job then you didn't do it! And presumably the cleaners only cleared waste paper bins and nothing else.

I think that these attitudes may have improved a bit but more recently (in 2009) I saw it again when a member of the Tasmania Parks and Environment Department, when asked to process our walking pass for the Overland Track said that we'd have to wait until later in the day "as it is more convenient for me then." She said this without a hint of embarrassment, even when I queried it by repeating her words. The customer certainly doesn't come first in these departments!

I don't know what the government managers did to their employees, maybe they bullied or brainwashed them, but certainly taken as a group they showed remarkable resistance to any idea of change. When demonstrating a technique, tool or process or something that was new or strange to them they would often respond with "but that's not how we do it," or "we wouldn't be allowed to do it that way." I would point out that we'd already established that things were not going well, that they'd been sent on the course by management to learn how to do things better. "If you keep on doing what you have been doing," I would say, "then you're going to get the same results that you've been getting." They would listen and shrug. I rarely experienced this sort of resistance in private enterprise. I am not surprised when I read (all too frequently) of yet another cock-up on a huge scale of a government IT project, where the waste runs into hundreds of millions of dollars. In 2013 it was reported that the Queensland Health Department spent over *one billion dollars* on software that was scrapped as it failed to do the job! A replacement was going to cost one and a third billion dollars. Similar stories are reported every year, many from the public sector, and the underlying causes have a certain deja vu about them, and they are nearly all basic issues!

One government department asked for training to help them with user system testing. A new system was about to be handed over to them, and they planned to carry out their own testing to ensure that all was well, following best practice. I ran two such courses which were highly successful, partly because the departmental director and his senior staff attended the training. As part of the training I used information about the new system in exercises about the planning of the tests, so that when the testing phase began they would be up and running from day one. The director reported back to me that the testing phase was a great success, saying that the training had been a crucial element. Some departments did get it right, and it was clear to me that this director made the difference – his staff were motivated, keen and eager as I had never seen before in a government department. Computer systems testing is an oft-neglected area as I have been reminded as I was writing this. A major bank in Australia has a huge problem with foreign exchange as their systems appear to have been used for large-scale money-laundering. Routine reporting checks that should have been installed have apparently never worked, it has been reported in the press! The most basic test should have shown this up during the testing phase.

We occasionally ran a free seminar for managers as a marketing device. At one of these in Canberra I had an audience of senior IT managers, many of them from defence. I was focusing on project management and I made the observation that in many organisations the project managers reported at a low level, and that consequently the bad news that should get reported was sifted out as the report went up the management levels, as managers never like (or are never encouraged) to report bad news. The result was that the senior management, the ones that could be taking action to fix things, were often not aware that anything was seriously amiss until very late in the day. I further suggested that this was easily fixed – by ensuring that the project manager reported to, and had easy access to, the top levels of management. A number of jaws all dropped in unison. Defence in particular pointed out that their project managers were generally at a lowly level and they couldn't possibly sit at the same table as the senior ranking staff officers! I replied that in private enterprise this often actually happened, and that if they continued to work as they had been working then… it later came as no surprise to me to learn that the software for the Collins Class submarines had cost five hundred thousand dollars to develop to date, but was way behind schedule and over budget… and was to be scrapped! More than a billion dollars was spent over the next couple of years to get software built in the USA, but not before the Defence Minister had told parliament in Canberra that he would buy the software off the shelf and have it working within a year! Who advises these people?

The lack of honest communication up the chain of command is very common in large organisations. At one financial organisation in Melbourne I ran a number of project management courses for their very large I.T. department, numbering in the thousands. I then presented a review of my findings to the senior management team, based on the delegate post-course reviews, (where they scored the course against a number of criteria) and a separate questionnaire on their working practices. I reported that a majority of the project managers felt that they were bullied to report only good news and to withhold bad news. The director was visibly astonished. He then asked his management team (a dozen or so senior managers) whether they thought that this could possibly be true. There was a shuffle of feet and a silence. Eventually one brave soul said that yes, he believed that it was true. Then the others all agreed with him. The director was dumbfounded. Whether the culture changed or not I can't say, but at least they knew, and could easily do something about it if they wanted to.

Another common and major issue was that senior managers would, without any knowledge of software development, set implementation dates in stone for a system that had yet to be specified, staffed and costed. A good project manager will attempt to talk about this, explain the issues and look at a more flexible and sensible approach. If the so-called "deadline" really cannot be moved (and most times it can), then the only way to deal with this realistically is to only specify that which can be handled within the time frame within the given resources. But this was never recognised and totally unrealistic situations quickly developed. Senior managers would not accept any news other than "we are on track" and would thump the desk, shout and bully, thinking presumably that this behaviour would spurn on the staff to achieve the impossible. It never did, of course, and another disaster would unfold. I've seen it happen so many times, yet always, if anyone had asked the people at the coal face what they thought, the real answer would have been emphatic and clear!

Project management courses gave me the most satisfaction of all the courses to present. Over the years I modified the course little by little and it became a very successful and popular course. Many of the improvements originated from the delegates, from whom of course I learnt much. And I read more books on the subject than ever before. And of course my experiences from Sainsbury, Plessey and the consulting projects went in as well. And, best of all, I occasionally managed to get some consulting from a business resulting from a successful project management course. The satisfaction that I derived came from the changes that I could make to the workings of a company's software development process. I would start the course by asking for examples of the delegates' projects that had gone wrong, and we would analyse them. The examples that came up were jaw-dropping. I heard of projects that cost millions being scrapped, never implemented, implemented but withdrawn... because of basic faults in the project management process. This information was given in confidence, and would have caused much embarrassment to the companies had it got out, as most of it had never been disclosed to the public. We would go through the course and at the end I would draw their attention to the issues that we'd identified and that we'd shown how the risks could be minimised or removed. When I ran the course in-house there was a much better chance that the necessary improvements would actually be made. And sometimes I had the opportunity to really help. Like when I ran courses for a Sydney software house that was experiencing serious issues. The director asked me to present the courses residentially, to my surprise, explaining that they had cultural issues to overcome (residential courses had never been popular in Australia, unlike the U.K., and then were as rare as hens' teeth). He warned me that I'd be facing entrenched opposition. I had always preferred the residential course from a training perspective, as the presenter had so much more control of the whole training environment, and the results were always assured. At the first course, held in the Blue Mountains, I faced the rows of senior staff, sitting with folded arms and hard looks. The director sat with them. By morning coffee they were warming, by lunchtime we were pals. The course was a great success and I ran a number more, and another of our technical courses, all residential. We changed the culture and their processes and the improvements were soon to be seen. Then I was asked to facilitate their annual get-together of their senior management teams from the USA and the UK, together with the Aussies. This was held in the Cotswolds in England and we were together in a delightful hotel for four days while I got on with the job I'd been assigned – to get the different managers working together and to break down the significant cultural differences that was causing them much bother! This was obvious on the first morning when they immediately began arguing over every little point! It was hard work and I had to use all of the tricks and techniques that I'd learnt over the years, but it was worth it when I was told that they all thought it had been a great success. One afternoon I had them all out on bicycles riding around Moreton-in-the-Marsh and playing bar skittles in the pub after and the Americans in particular enjoyed themselves immensely. Of course I combined this all-expenses-paid trip with a trip to Clacton, London and Edinburgh! The Sydney-based director who had put me forward for this work said afterwards how pleased the directors had all been and that the business was showing great improvements now that the management teams understood each other.

My business had some ups and downs over the years as we weathered the financial storms and some storms of our own making. We had a number of staff come and go, and I was not very successful with choosing salespeople. Until the last one, that is, when Gina was involved with the selection and we chose a fellow who later became the general manager, enabling me to retire! I did much of the marketing and selling, which I enjoyed, and I was good at it as I knew the courses intimately and I knew what issues the managers faced and how the training would help them and their staff. I quite enjoyed establishing contacts and bringing a client to the buying point. Once I got to the right person – the decision maker – I rarely lost a sale. But I was also doing much of the course presenting, so we required a salesperson as well. One sale I made was through a competitive tender for a Sydney financial institution. It was quite a process and I had to make a final presentation to a group of their managers. We won, and it resulted in a string of courses that went on for ten years! We had several receptionists/bookkeepers over the years, and of course I was always on the lookout for good presenters. We usually had four or five on the books and we'd allocate jobs to them to suit their skills and experience, as they couldn't all run every course. I had to train them and produce a presenter's guide for each course. I had a few issues to deal with when something went wrong with a course presentation, but these were thankfully few but I had to drop a couple of presenters. Our best presenter was a very likeable, extremely bright chap who was a funny and engaging speaker. Unlike the other presenters, he took up a full-time job with us for a while. Tragically he succumbed to a brain tumour and died in his fifties.

Balancing work with bushwalking became an important issue. I would often fly back to Melbourne on a Friday evening to be met by Gina at the airport with all the camping/walking gear in the car and we'd drive straight off to the rendevous campsite ready for a weekend bushwalk. Or we'd finish a bushwalk on Sunday, drive to the airport so I could catch my flight to Sydney or wherever. Sometimes I would be on the plane still in my walking gear, somewhat sticky and probably a bit on the nose! This involved being very organised of course, with packs, suitcases, course materials ready packed a week or so in advance. Gina and I became very involved with the bushwalking club and we served on the management committee in various roles over ten years or so, including Gina being the walks secretary and myself becoming president. We joined a second club and again became involved with the management of it.

We've made some very good friends in the clubs and we've taken part in so many adventures with them, both in club trips and in small private groups.

During this time I became a member of the Victorian Board of the Australian Computer Society, which I had always felt had never reached its potential. One of my duties was to talk to university students about I.T. jobs. After one of the economic dips in the Nineties there was a lull in I.T. job vacancies and a severe drop in the number of uni applicants for the technology courses. I spoke to several undergraduate groups about the job market and tried to impress upon them that what we were seeing was just a temporary lull in the job market. I told them how much fun I had had in my career and that computers were so ingrained now in every aspect of life that it was one of the most secure professions to be in. I hope that I was able to encourage them to stick with their technology courses, there were going to be plenty of opportunities for them. I retired from the board when I retired from the business and I'm pleased to see that it is still going well.

My business partner started off being the book-keeper, at which he proved to be less than adequate! At first he would write cheques and leave the sub blank so we had no record of what we'd paid, for what and to whom! I had to remonstrate loudly to stop this. I took over the book-keeping for a time until we employed someone to do this. He was happier as the marketing man and he handled much of our mail-out operations (although I wrote much of the material). He looked after the office infrastructure and acquired the P.C. s, network, software, and kept us up-to-date as the internet arrived with emailing etc. He embraced the internet and built our first website, tending to try all the gimmicks. I had little interest in this side of things. He was a good talker and a great ideas man, and he was a good and energetic starter. He was fascinated by a technical challenge and would beaver away to make it work on a technical level, but he was not so keen on the follow-through. He was always looking for another way to make money, while I focused on what I thought we were good at – training courses for business analysts. He started up many projects, most of which failed to go anywhere. I felt guilty and negative as I put down his ideas but he would immediately bounce back with another. His major success was with multimedia – producing digitised CDs and DVDs for training and promotion, from a client's own material. For a time this went well, and reached the point where multimedia generated forty per cent of our total revenue. We had a contract with Ford, and another with the Victorian Education Department, and we sent three chaps to the USA to work on a contract there. To achieve this we had to invest a lot of cash in hardware and software in order to remain on the leading edge. Some of the hardware, for example, would cost tens of thousands of dollars, and a software product once cost us fifty thousand dollars and was available for a few thousand just a few months later! I insisted that we split the business, so that we could see the costs and revenues of the two sides of the business separately as I could see that training could be subsidising multimedia. We then made multimedia a separate business entity later.

During this time I was always going off on bushwalking trips, mainly weekends but once or twice a year we'd do a trip of several weeks. After one such trip I got back to the office to find that my partner had taken on four telephone sales callers. He'd decided that telephone cold calling was the way to go and he'd recruited four people, none of whom had any knowledge of IT, yet alone the type of training that we did. The operation failed completely after the senior operator had upset the other staff, and no additional sales were forthcoming. I had the job of closing it down and sacking the staff; my partner couldn't or wouldn't do this. There were several other similar catastrophes awaiting for me on my return from other trips. Although we exchanged strong words he would bounce back quickly with the next idea...

I continued in the business until I felt that I'd had enough. This happened quite quickly, when after the millennium issue (when every computer programme in use had to be checked to ensure that it could handle the date changeover) and the implementation of GST I realised that the issues arising had all been seen and dealt with in some form before, and although I was still enjoying presenting courses myself I no longer felt the kick from fixing things. And my hearing was getting worse so I was finding it more difficult to get rapport with the delegates. My partner wished to retire as well so there were three options I could take with the business: close it, sell it, or get someone else to run it.

I began to draw back from presenting, and made plans so that the business could continue without me. It took almost eighteen months. Some things that I'd been doing had to be formalised. Gina had been working with me for several years as the administrator and bookkeeper (the best that we'd had) and the booking and financial systems had to be set up so that someone could take over, as I wanted her to retire with me, naturally. At Christmas 2002, when I was sixty one, I formally retired. Our salesman had been groomed, he had my trust and he took over as general manager. I'd worked out a contract with him whereby he had the opportunity of taking shares in the company if he met certain targets. Over the following years he sometimes made the targets, sometimes he didn't. In this way he acquired shares and soon owned a chunk of the business. Gina stayed on for a short time to assist in the handover, and train up the new administrator, who turned out to be a gem.

Gina and I were now free to enjoy our retirement and follow our interests, and that was going to focus on travel and bushwalking. In 2004 we toured Australia in our 4WD for seven months and did another three-month tour the next year. Many adventures followed. For the first five years I ran a course or two each year to help out when required. In 2005 when our key presenter was taken ill with the brain tumour I had to jump back in and run courses again for some months until we found replacement presenters. The last course I ran was for the RAAF in 2006 and that was a technical writing course. It went well, I enjoyed it but thought then that it was probably the last one. It was.

POSTSCRIPT

While writing this I have referred to my (inadequate) notes, diaries and papers. It has been a sporadic affair, with often nothing happening for months, then a paragraph or two added on some whim. While writing the chapter about my Sainsbury experiences I was prompted to make a contribution to the Sainsbury archives, held at the Museum of London, Docklands, that one day I have promised to visit. Now, in 2017 I reflect that one of my Keith London partners, Colin Corder is no longer with us – he dropped dead at home from a heart attack in 1994. Colin was a very clever man, an author and writer *par excellence*, a mover and shaker, a delightful man to know, and from whom I learnt much. Another, Tony Elkington, the most energetic KLA salesman who was always full of good ideas, suddenly disappeared from the radar entirely sometime during the early 2000s. I keep some contact with a couple of the other partners and I've had contact with a couple of my ex-staff from Plessey days. I also talk to some buddies from the Boy Scouts and from my college days. Otherwise, one's past rapidly disappears from view, and that is probably as well. The four factories where I worked – Hoffmann in Chelmsford, Electrical Apparatus Co in St Albans, Marconi Instruments, St Albans, and the Plessey factory at Ilford have all been levelled and replaced with housing estates, and the companies no longer exist, reflecting the manufacturing decline in the UK and the ever-increasing UK population. The Sainsbury buildings at Blackfriars have also gone, and the Sainsbury head office has relocated and is now split up all across England. The business that I started in Melbourne is still an operating business and I remained a director until we sold the business in 2016. This is now the only sign that I ever existed from the work viewpoint, except for the barcodes on the Sainsbury shelves! Perhaps I'm forgetting that out there somewhere are several thousand people who went on my training courses but probably few would remember me or the courses. I just hope that the courses may have helped their careers a little.

Is that much to show for a lifetime's effort? I suppose that it is more than my parents ever expected, and more than I could ever have imagined in my youth.

Now, in 2017, I am enjoying retirement with Gina who has been and is a great comfort and joy to me after some thirty years together. My daughter Lucy and husband Mike have given me a wonderful grand-daughter, Emi, now aged nine, an absolute delight and we get together every year for a holiday. They live in Edinburgh, much too far away!

I began to write this some years ago thinking only of my descendants and extended family, then it just grew. If it has been interesting in some way, I'm pleased. I've tried to ensure that dates, places and events are accurate but of course much of this is subject to my fallible memory. I have been a little free with the chronology here and there in order to assist the flow. Any factual mistakes are mine.

Melbourne, August 2017